A Brief History of Computing

Gerard O'Regan

A Brief History of Computing

Springer

Gerard O'Regan, BSc, MSc, PhD
11 White Oaks, Mallow, Co. Cork, Ireland
oregang@yahoo.com

ISBN: 978-1-84800-083-4 e-ISBN: 978-1-84800-084-1
DOI: 10.1007/978-1-84800-084-1

British Library Cataloguing in Publication Data
A catalogue record for this book is available from the British Library

Library of Congress Control Number: 2007941396

Printed on acid-free paper

9 8 7 6 5 4 3 2 1

Springer Science+Business Media
springer.com

To my wonderful nieces and nephews Jamie,
Tiernan, Cian, Aoife, Lorna and Daniel

Preface

Overview

The objective of this book is to provide an introduction into some of the key topics in the history of computing. The computing field is a vast area and a truly comprehensive account of its history would require several volumes. The aims of this book are more modest, and its goals are to give the reader a flavour of some of the key topics and events in the history of computing. It is hoped that this will stimulate the interested reader to study the more advanced books and articles available.

The history of computing has its origins in the dawn of civilization. Early hunter gatherer societies needed to be able to perform elementary calculations such as counting and arithmetic. As societies evolved into towns and communities there was a need for more sophisticated calculations. This included primitive accounting to determine the appropriate taxation to be levied as well as the development of geometry to enable buildings, templates and bridges to be constructed. Our account commences with the contributions of the Egyptians, and Babylonians. It moves on to the foundational work done by Boole and Babbage in the nineteenth century, and to the important work on Boolean Logic and circuit design done by Claude Shannon in the 1930s. The theoretical work done by Turing on computability is considered as well as work done by von Neumann and others on the fundamental architecture for computers.

The various generations of programming languages that have been developed over the last 50 years are then considered. This includes machine code used in the early days of programming; low-level assembly languages; high-level programming languages such as C and Pascal, and so on. The important field of software engineering is then considered, and this chapter considers various software engineering methodologies to design and develop high-quality software to meet customers' requirements. This includes a discussion on theoretical mathematical approaches as well as practical industrial approaches using the CMMI.

The field of Artificial Intelligence is then considered starting with Greek mythology on creating life and intelligence. The Turing test developed by Alan Turing to judge whether a machine may be considered intelligent is considered as well as Searle's rebuttal argument on Artificial Intelligence.

Next, the development of the internet and world-wide web is considered. This includes the work done by DARPA on the internet as well as the work done by Tim Berners-Lee at CERN that led to the birth of the world-wide web. Applications of the world-wide web are considered as well as the dot com bubble and eventual collapse.

Finally, the contributions of various well-known technology companies are considered. These include companies such as International Business Machines (IBM), Microsoft, and Motorola.

Organization and Features

The first chapter discusses the contributions made by early civilizations to computation. This includes work done by the Babylonians, Egyptians and Greeks. The Babylonians recorded their mathematics on soft clay which they then baked in ovens. The Egyptians applied their mathematics to solving practical problems including the construction of pyramids as well as various accounting problems. The Rhind Papyrus is one of the most famous Egyptian papyri on mathematics, and it is essentially a mathematical text book. The Greeks made a major contribution to mathematics and geometry, and most students are familiar with the work of Euclid on geometry.

Chapter two considers influential figures in the history of computing who did important foundational work. It includes a discussion on Boole, Babbage, Turing, Shannon, and von Neumann. Boolean logic is fundamental to the working of all modern computers, and Boole is considered to be one of the fathers of computing. Babbage did pioneering work on the Difference Engine as well as designing the Analytic Engine. The difference engine was essentially a mechanical calculator while the analytic engine was essentially the world's first computer. However, the analytic engine was never built. Lady Ada Lovelace who was a friend of Babbage's designed the first programs to run on the analytic engine, and she believed that the machine would be applicable to many disciplines.

Claude Shannon showed how Boolean Logic could be applied to simplify the design of circuits, and how Boolean logic could be employed in computing. Turing's work on computability showed that whatever was computable was computable by his theoretical Turing Machine, and he also made contributions to the Artificial Intelligence field. Von Neumann and others defined the von Neumann architecture which is the fundamental architecture used in modern computers.

Chapter three considers various programming languages developed over the last fifty years. These include the five generations of programming languages such as machine code languages that use actual machine instructions, to assembly languages, to high-level programming languages such as Cobol, Fortran, Pascal, C, C++ and Java. The earliest high-level language developed was called Plankalkül, and this language was developed by the German engineer Zuse. Functional programming languages are discussed as well as the important area of syntax and semantics of programming languages.

Chapter four considers the important field of software engineering. It discusses the birth of software engineering at the NATO conference at Garmisch in 1968. This conference discussed the crisis with software and included problems with projects being delivered late or with poor quality. The software engineering field is concerned with sound techniques to engineer high quality software to meet customers' requirements. It includes methodologies to define requirements, design, development and test software as well as the management of change. Mathematical techniques that may assist in software engineering are also considered, and these include approaches such as Z and VDM. The Capability Maturity Model Integration model is discussed, as this offers a useful framework to enhance the software engineering capability of a company.

Chapter five considers Artificial Intelligence and Expert Systems. Artificial Intelligence is a multi-disciplinary field that is concerned with the problem of producing intelligence in machines. This chapter considers some of the key disciplines in the field including philosophy, psychology, linguistics, neural networks, and so on. The Turing Test was proposed by Alan Turing to judge whether a machine is intelligent. Searles's Chinese room argument is a rebuttal and argues that even if a machine passes the Turing Test it still may not be considered intelligent.

Chapter six is concerned with the Internet revolution and considers the early work done by DARPA on the internet, and the subsequent invention of the world-wide web by Tim Berners-Leee at CERN in the early 1990. The rise of new technology companies such as Amazon is considered, as well as the frenzy that became the dot com bubble and subsequent dot com collapse. Many of these new technology companies had deeply flawed business models which was a significant factor in their eventual collapse. Several of dot com failures are considered including the on-line fashion company Boo.com and also the on-line pet food company Pets.com.

Chapter seven is concerned with the achievements of several of the better known software technology companies. These include IBM, Motorola and Microsoft.

Audience

This book is suitable for students of Computer Science and for the general reader who is interested in the history of computing. Some of the material is technical and mathematical. However, any technical material is presented as simply as possible.

Acknowledgments

I would like to express thanks to IBM Archives for permission to include photographs of the IBM PC, IBM 360, Hermann Hollerith, Hollerith's Tabulating Machine from 1890, the photo of the Deep Blue processors, as well as photos of Thomas Watson Senior and Junior. I must thank Brian Randell for permission to use the photo of Dijkstra from the 1968 software engineering conference. I would like to thank the W. Edwards Deming Institute for permission to use a photo of Deming and the Juran Institute for permission to use a photograph of Juran. I must thank Fred Brooks, Watt Humphries and John McCarthy for permission to use their photographs. I must thank Horst Zuse for the use of a photograph of Konrad Zuse. I would also like to thank the School of Computing at the University of Manchester for permission to include a photograph of the Manchester Mark 1. The National Physical Laboratory gave permission to reproduce the NPL ACE machine from 1950. I would like to thank Tommy Thomas for the use of a photo of the replica of the Manchester Mark 1 Baby that was taken at the Manchester Museum of Science and Industry. I would like to thank Wikipedia for using its public domain photographs.

I am deeply indebted to family, friends and colleagues in industry and academia who supported my efforts in this endeavour. A special thanks to Mary Rose, James and Jamie for their company and hospitality in Dublin. My thanks to friends in the Cork area especially Kevin, Noel, Maura (x2), and Gerry. Finally, I must thank the team at Springer especially Catherine Brett, Wayne Wheeler and the production team. Finally, my thanks to the reviewers for their improvement suggestions and for helpful comments.

Gerard O'Regan
Cork, Ireland

Contents

List of Figures

List of Tables

Chapter 1
Early Civilisations

Key Topics

Babylonian Mathematics
Egyptian Civilisation
Greek and Roman Civilisation
Counting and Numbers
Solving Practical Problems
Syllogistic Logic
Algorithms
Early Ciphers

1.1 Introduction

It is difficult to think of western society today without modern technology. The last decades of the twentieth century have witnessed a proliferation of high-tech computers, mobile phones, text-messaging, the internet and the world-wide web. Software is now pervasive and it is an integral part of automobiles, airplanes, televisions, and mobile communication. The pace of change as a result of all this new technology has been extraordinary. Today consumers may book flights over the world-wide web as well as keeping in contact with family members in any part of the world via email or mobile phone. In previous generations, communication often involved writing letters that took months to reach the recipient. Communication improved with the telegrams and the telephone in the late nineteenth century. Communication today is instantaneous with text-messaging, mobile phones and email, and the new generation probably views the world of their parents and grandparents as being old-fashioned.

The new technologies have led to major benefits[1] to society and to improvements in the standard of living for many citizens in the western world. It has also reduced

[1] Of course, while the new technologies are of major benefit to society it is essential that the population of the world moves towards more sustainable development to ensure the long-term survival of the planet for future generations. This involves finding technological and other solutions

the necessity for humans to perform some of the more tedious or dangerous manual tasks, as many of these may now be automated by computers. The increase in productivity due to the more advanced computerized technologies has allowed humans, at least in theory, the freedom to engage in more creative and rewarding tasks.

This chapter considers work done on computation by early civilizations including the work done by our ancestors in providing a primitive foundation for what has become computer science. Early societies are discussed and their contributions to the computing field are considered. There is a close relationship between the technological or computation maturity of a civilization and the sophistication of its language. Clearly, societies that have evolved technically will have invented words to reflect the technology that they use on a daily basis. Clearly, hunter-gatherer or purely agrarian societies will have a more limited technical vocabulary which reflects that technology is not part of their day to day culture. Language evolves with the development of the civilisation, and new words are introduced to describe new inventions in the society. Communities that have a very stable unchanging existence (e.g., hunter gatherer or pastoral societies) have no need to introduce names for complex scientific entities, as these words are outside their day-to-day experience. The language of these communities mirrors the thought processes of these communities.

Early societies had a limited vocabulary for counting: e.g., "one, two, three, many" is associated with some primitive societies, and indicates primitive computation and scientific ability. It suggests that there was no need for more sophisticated arithmetic in the primitive culture as the problems dealt with were elementary. These early societies would typically have employed their fingers for counting, and as humans have 5 fingers on each hand and five toes on each foot then the obvious bases would have been 5, 10 and 20. Traces of the earlier use of the base 20 system are still apparent in modern languages such as English and French. This includes phrases such as "three score" in English and "*quatre vingt*" in French.

The decimal system (base 10) is familiar to most in western society, and it may come as a surprise that the use of base 60 was common in computation *circa* 1500 BC. One example of the use of base 60 today is still evident in the sub-division of hours into 60 minutes, and the sub-division of minutes into 60 seconds. The base 60 system (i.e. the sexagesimal system) is inherited from the Babylonians [Res:84], and the Babylonians were able to represent arbitrarily large numbers or fractions with just two symbols. Other bases that have been used in modern times include binary (base 2) and hexadecimal (base 16). Binary and hexadecimal arithmetic play a key role in computing, as the machine instructions that computing machines understand are in binary code.

The ancient societies considered in this chapter include the Babylonians, the Egyptians, and the Greek and Romans. These early civilizations were concerned with the solution of practical problems such as counting, basic book keeping, the construction of buildings, calendars and elementary astronomy. They used

---

to reduce greenhouse gas emissions as well as moving to a carbon neutral way of life. The solution to the environmental issues will be a major challenge for the twenty first century.

appropriate mathematics to assist them in computation. Early societies had no currency notes like U.S. Dollars or Euros, and trading between communities was conducted by bartering. This involved the exchange of goods for other goods at a negotiated barter rate between the parties. This required elementary computation as the bartering of one cow would require the ability to agree that a cow was worth so many of another animal, crop or good, e.g., sheep, corn, and so on. Once this bartering rate was agreed the two parties then needed to verify that the correct number of goods was received in exchange. Therefore, the ability to count was fundamental.

The achievements of some of these ancient societies were spectacular. The archaeological remains of ancient Egypt are very impressive, and include the pyramids at Giza, the temples of Karnak near Luxor and Abu Simbal on the banks of Lake Nasser. These monuments provide an indication of the engineering sophistication of the ancient Egyptian civilisation. The objects found in the tomb of Tutankamun[2] are now displayed in the Egyptian museum in Cairo, and demonstrate the artistic skill of the Egyptians.

The Greeks made major contributions to western civilization including contributions to Mathematics, Philosophy, Logic, Drama, Architecture, Biology and Democracy.[3] The Greek philosophers considered fundamental questions such as ethics, the nature of being, how to live a good life, and the nature of justice and politics. The Greek philosophers include Parmenides, Heraclitus, Socrates, Plato and Aristotle. The works of Plato and Aristotle remain important in philosophy today, and are studied widely. The Greeks invented democracy and their democracy was radically different from today's representative democracy.[4] The sophistication of

[2] Tutankamun was a minor Egyptian pharaoh who reigned after the controversial rule of Akenaten. Tutankamun's tomb was discovered by Howard Carter in the valley of the kings, and the tomb was intact. The quality of the workmanship of the artefacts found in the tomb was extraordinary and a visit to the Egyptian museum in Cairo is memorable.

[3] The origin of the word "democracy" is from demos (δημος) meaning people and kratos (κρατος) meaning rule. That is, it means rule by the people. It was introduced into Athens following the reforms introduced by Cleisthenes. He divided the Athenian city state into thirty areas. Twenty of these areas were inland or along the coast and ten were in Attica itself. Fishermen lived mainly in the ten coastal areas; farmers in the ten inland areas; and various tradesmen in Attica. Cleisthenes introduced ten new clans where the members of each clan came from one coastal area, one inland area on one area in Attica. He then introduced a Boule (or assembly) which consisted of 500 members (50 from each clan). Each clan ruled for $^{1}/_{10}$th of the year.

[4] The Athenian democracy involved the full participations of the citizens (i.e., the male adult members of the city state who were not slaves) whereas in representative democracy the citizens elect representatives to rule and represent their interests. The Athenian democracy was chaotic and could also be easily influenced by individuals who were skilled in rhetoric. There were teachers (known as the Sophists) who taught wealthy citizens rhetoric in return for a fee. The origin of the word "sophist" is the Greek word σοφος meaning wisdom. One of the most well known of the sophists was Protagorus, and Plato has a dialogue of this name. The problems with the Athenian democracy led philosophers such as Plato to consider alternate solutions such as rule by philosopher kings. This is described in Plato's Republic.

Greek architecture and sculpture is evident from the Parthenon on the Acropolis, and the Elgin marbles[5] that are housed today in the British Museum, London.

The Hellenistic[6] period commenced with Alexander the Great and led to the spread of Greek culture throughout Asia Minor and as far as Egypt. The city of Alexandria was founded by Alexander the Great, and it became a center of learning and knowledge during the Hellenistic period. Among the well-known scholars at Alexandria was Euclid who provided a systematic foundation for geometry. Euclid's work on geometry is known as "The Elements", and it consists of thirteen books. The early books are concerned with the construction of geometric figures, number theory and solid geometry.

There are many words of Greek origin that are part of the English language. These include words such as psychology which is derived from two Greek words: *psyche* (ψυχε) and *logos* (λογος). The Greek word "*psyche*" means mind or soul, and the word "*logos*" means an account or discourse. Other examples are anthropology derived from "*anthropos*" (αντροπος) and "*logos*" (λογος).

The Romans were influenced by the culture of the Greeks, and the Greek language, culture and philosophy was taught in Rome and in the wider Roman Empire. The Romans were great builders and their contributions include the construction of buildings such as aqueducts, viaducts, baths, and amphitheatres. Other achievements include the Julian calendar, the formulation of laws (*lex*), and the maintenance of law and order and peace throughout the Roman Empire. The peace that the Romans brought is known as *pax Romano*. The ruins of Pompeii and Herculaneum (both located near Naples in Italy) demonstrate the engineering maturity of the Roman Empire. The Roman numbering system is still employed today in clocks and for page numbering in documents. However, as a notation it is cumbersome for serious computation. The collapse of the Roman Empire in Western Europe led to a decline in knowledge and learning in Europe. However, the eastern part of the Roman Empire continued at Constantinople (now known as Istanbul in Turkey) until its sacking by the Ottomans in 1453.

1.2 The Babylonians

The Babylonian[7] civilization flourished in Mesopotamia (in modern Iraq) from about 2000 BC until about 300 BC. Various clay cuneiform tablets containing mathematical texts were discovered and later deciphered by Grotefend and Rawlinson in the nineteenth century [Smi:23]. These included tables for multiplication, division, squares, cubes and square roots; measurement of area and length; and the solution

[5] The Elgin marbles are named after Lord Elgin who moved them from the Parthenon in Athens to London in 1806. The marbles show the Pan-Athenaic festival that was held in Athens in honour of the goddess Athena after whom Athens is named.

[6] The origin of the word Hellenistic is from Hellene ("Ελληυ") meaning Greek.

[7] The hanging gardens of Babylon were one of the seven wonders of the ancient world.

of linear and quadratic equations. The late Babylonian period (c. 300 BC) includes work on astronomy.

The Babylonians recorded their mathematics on soft clay using a wedge shaped instrument to form impressions of the *cuneiform* numbers. The clay tablets were then baked in an oven or by the heat of the sun. They employed just two symbols (1 and 10) to represent numbers, and these symbols were then combined to form all other numbers. They employed a positional number system[8] and used base 60 system. The symbol representing 1 could also (depending on the context) represent 60, 60^2, 60^3, etc. It could also mean 1/60, 1/3600, and so on. There was no zero employed in the system and there was no decimal point (strictly speaking no "sexagesimal point"), and therefore the context was essential.

The example above illustrates the cuneiform notation and represents the number $60 + 10 + 1 = 71$. The Babylonians used the base 60 system for computation, and this base is still in use today in the division of hours into minutes and the division of minutes into seconds. One possible explanation for the use of base 60 is the ease of dividing 60 into parts as it is divisible by 2, 3, 4, 5, 6, 10, 12, 15, 20, and 30. They were able to represent large and small numbers and had no difficulty in working with fractions (in base 60) and in multiplying fractions. They maintained tables of reciprocals (i.e., $1/n$, $n = 1, \ldots 59$ apart from numbers like 7, 11, etc., which are not of the form $2^{\alpha}3^{\beta}5^{\gamma}$ and cannot be written as a finite sexagesimal expansion).

Babylonian numbers are represented in the more modern sexagesimal notation developed by Neugebauer (who translated many of the Babylonian cuneiforms) [Res:84]. The approach is as follows: 1;24,51,10 represents the number $1 + 24/60 + 51/3600 + 10/216000 = 1 + 0.4 + 0.0141666 + 0.0000462 = 1.4142129$ and is the Babylonian representation of the square root of 2. The Babylonians performed multiplication as the following calculation of (20) * (1; 24, 51, 10) i.e., 20 * sqrt(2) illustrates:

$$20^{*}1 = 20$$

$$20^{*}; 24 = 20^{*}\frac{24}{60} = 8$$

$$20^{*}\frac{51}{3600} = \frac{51}{180} = \frac{17}{60} =; 17$$

$$20^{*}\frac{10}{216000} = \frac{3}{3600} + \frac{20}{216000} =; 0, 3, 20$$

Hence the product 20 * sqrt(2) = $20; +8; +; 17+; 0, 3, 20 = 28; 17, 3, 20$

The Babylonians appear to have been aware of Pythagoras's Theorem about 1000 years before the time of Pythagoras. The Plimpton 322 tablet records various

---

[8] A positional numbering system is a number system where each position is related to the next by a constant multiplier. The decimal system is an example: e.g., $546 = 5^{*}10^2 + 4^{*}10^1 + 6$.

Fig. 1.1 The plimpton 322 tablet

Pythagorean triples, i.e., triples of numbers (a, b, c) where $a^2 + b^2 = c^2$ (Fig. 1.1). It dates from approximately 1700 BC

They developed algebra to assist with problem solving, and their algebra allowed problems involving length, breadth and area to be discussed and solved. They did not employ notation for representation of unknown values (e.g., let x be the length and y be the breadth), and instead they used words like "length" and "breadth". They were familiar with and used square roots in their calculations, and while they were familiar with techniques to solve quadratic equations.

They were familiar with various mathematical identities such as $(a+b)^2 = (a^2 + 2ab + b^2)$ as illustrated geometrically in Fig. 1.2. They also worked on astronomical problems, and they had mathematical theories of the cosmos to make predictions of when eclipses and other astronomical events would occur. They were also interested in astrology, and they associated various deities with the heavenly bodies such as the planets, as well as the sun and moon. Various cluster of stars with associated with familiar creatures such as lions, goats, and so on.

The earliest form of counting by the Babylonians was done using fingers. They improved upon this by developing counting boards to assist with counting and

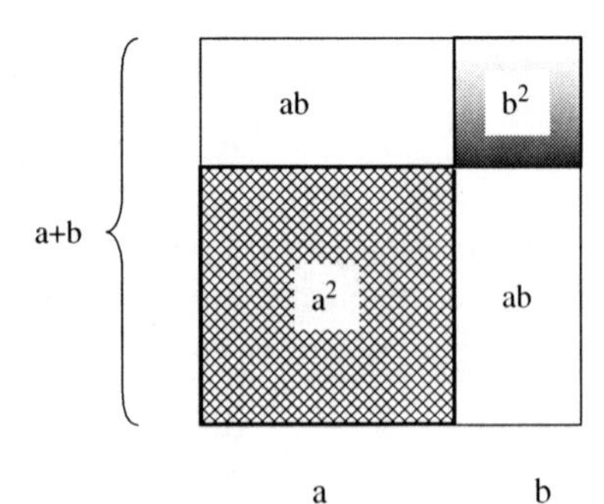

Fig. 1.2 Geometric representation of $(a + b)^2 = (a^2 + 2ab + b^2)$

simple calculations. A counting board is an early version of the abacus, and it was usually made of wood or stone. The counting board contained grooves which allowed beads or stones could be moved along the groove. The abacus differed from counting boards in that the beads in abaci contained holes that enabled them to be placed in a particular rod of the abacus.

1.3 The Egyptians

The Egyptian Civilization developed along the Nile from about 4000 BC and lasted until the Roman Empire. The achievements of the Egyptian civilization are remarkable and their engineers built the gigantic pyramids at Giza near Cairo around 3000 BC.

The Egyptians used mathematics to solve practical problems. This included measuring time, measuring the annual Nile flooding, calculating the area of land, solving cooking and baking problems, book keeping and accounting, and calculating taxes. They developed a calendar circa 4000 BC. It consisted of 12 months, and each month had 30 days. There were then five extra feast days to give 365 days in a year. Egyptians writings were recorded on the walls of temples and tombs[9] and were also recorded on a reed like parchment termed "papyrus". There are three well-known Egyptian scripts namely the well-known hieroglyphics writing system; the hieratic script; and the demotic script.

The deciphering of the Egyptian hieroglyphics was done by Champollion with his work on the Rosetta stone. The latter was discovered during the Napoleonic campaign in Egypt, and is now in the British Museum in London. It contains three scripts: Hieroglyphics, Demotic script and Greek. The key to the decipherment was that the Rosetta stone contained just one name "Ptolemy" in the Greek text, and this was identified with the hieroglyphic characters in the cartouche[10] of the hieroglyphics. There was just one cartouche on the Rosetta stone, and Champollion inferred that the cartouche represented the name "Ptolemy". He was familiar with another multi-lingual object which contained two names in the cartouche. One he recognised as Ptolemy and the other he deduced from the Greek text as "Cleopatra". This led to the breakthrough in translation of the hieroglyphics [Res:84].

The Egyptians writing system is based on hieroglyphs and dates from 3000 BC. Hieroglyphs are little pictures and are used to represent words, alphabetic characters as well as syllables or sounds.

The Rhind Papyrus is one of the most famous Egyptian papyri on mathematics. It was purchased by the Scottish Egyptologist, Henry Rhind, in 1858 and is now in the British museum. The papyrus is a copy and it was created by an Egyptian

---

[9] The decorations of the tombs in the Valley of the Kings record the life of the pharaoh including his exploits and successes in battle.

[10] The cartouche surrounded a group of hieroglyphic symbols enclosed by an oval shape. Champollion's insight was that the group of hieroglyphic symbols represented the name of the Ptolemaic pharaoh "Ptolemy".

scribe called Ahmose.[11] It was originally six meters in length, and it is believed to date to 1832 BC. It contains examples of all kinds of arithmetic and geometric problems, and it was probably intended to be used by students as a textbook to develop their mathematical knowledge. This would allow the students to participate in the pharaoh's building programme. There is another well known papyrus known as the Moscow papryrus.

The Egyptian priests had were familiar with geometry, arithmetic and elementary algebra. They had formulae to find solutions to problems with one or two unknowns. A bases 10 number system was employed separate with symbols for one, ten, a hundred, a thousand, a ten thousand, a hundred thousand, and so on. These hieroglyphic symbols are represented in Fig. 1.3 below:

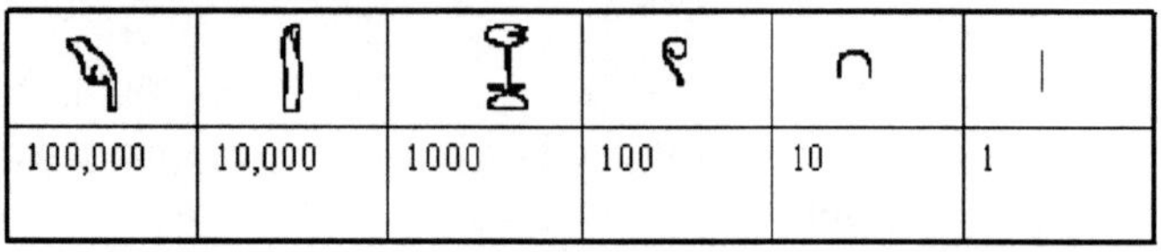

100,000	10,000	1000	100	10	1

Fig. 1.3 Egyptian numerals

For example, the representation of the number 276 in Egyptian Hieroglyphics is given in Fig. 1.4:

Fig. 1.4 Egyptian representation of a number

The addition of two numerals is straight forward and involves adding the individual symbols, and where there are ten copies of a symbol it is then replaced by a single symbol of the next higher value. The Egyptian employed unit fractions (e.g., $1/n$ where n is an integer). These were represented in hieroglyphs by placing the symbol representing a "mouth" above the number. The symbol "mouth" represents part of. For example, the representation of the number $1/276$ is given in Fig. 1.5:

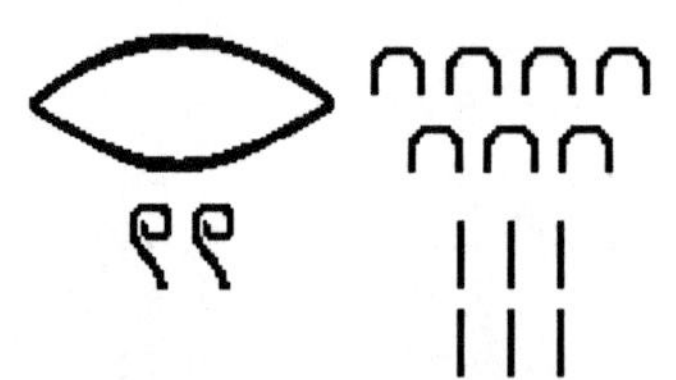

Fig. 1.5 Egyptian representation of a fraction

The problems on the papyrus included the determination of the angle of the slope of the pyramid's face. The Egyptians were familiar with trigonometry including sine, cosine, tangent and cotangent. They knew how to build right angles into

[11] The Rhind papyrus is sometimes referred to as the Ahmes papyrus in honour of the scribe who wrote it in 1832 BC.

their structures and used the ratio 3:4:5. The Rhind Papyrus also considers practical problems such as how many loaves of bread can be baked from a given quantity of grain. Other problems included calculating the number of bricks required for part of a building project. The Egyptians were familiar with addition, subtraction, multiplication and division. However, their multiplication and division was cumbersome as they could only multiply and divide by two.

Suppose they wished to multiply a number n by 7. Then $n * 7$ is determined by $n * 2 + n * 2 + n * 2 + n$. Similarly, if they wished to divide 27 by 7 they would note that $7 * 2 + 7 = 21$ and that $27 - 21 = 6$ and that therefore the answer was $3^6/_7$. Egyptian mathematics was cumbersome and the writing of their mathematics was long and repetitive. For example, they wrote a number such as 22 by $10+10+1+1$.

The Egyptians calculated the approximate area of a circle by calculating the area of a square 8/9 of the diameter of a circle. That is, instead of calculating the area in terms of our familiar πr^2 their approximate calculation yielded $(8/9 * 2r)^2 = 256/81\, r^2$ or $3.16\, r^2$. Their approximation of π was 256/81 or 3.16. They were able to calculate the area of a triangle and volumes. The Moscow papyrus includes a problem to calculate the volume of the frustum. The formula for the volume of a frustum of a square pyramid[12] was given by $V = 1/3\, h(b_1^2 + b_1 b_2 + b_2^2)$ and when b_2 is 0 then the well-known formula for the volume of a pyramid is given: i.e., $1/3\, hb_1^2$.

1.4 The Greeks

The Greeks made major contributions to western civilization including mathematics, logic, astronomy, philosophy, politics, drama, and architecture. The Greek world of 500 BC consisted of several independent city states such as Athens and Sparta, and various city states in Asia Minor. The Greek polis (πολισ) or city state tended to be quite small, and consisted of the Greek city and a certain amount of territory outside the city state. Each city state had political structures for its citizens, and these varied from city state to city state. Some were oligarchs where political power was maintained in the hands of a few individuals or aristocratic families. Others were ruled by tyrants (or sole rulers) who sometimes took power by force, but who often had a lot of support from the public. The tyrants included people such as Solon, Peisistratus and Cleisthenes in Athens.

The reforms by Cleisthenes led to the introduction of the Athenian democracy. This was the world's first democracy, and power was fully placed in the hands of the citizens. The citizens were male members of the population, and women or slaves did not participate. The form of democracy in ancient Athens differed from the representative democracy that we are familiar with today. It was an extremely liberal democracy where citizens voted on all important issues. Often, this led to disastrous results as speakers who were skilled in rhetoric could exert significant influence on the decision making. Philosophers such as Plato were against democracy as a form

[12] The length of a side of the bottom base of the pyramid is b_1 and the length of a side of the top base is b_2

of government for the state, and Plato's later political thinking advocated rule by philosopher kings. These rulers were required to study philosophy and mathematics for many years.

The rise of Macedonia led to the Greek city states being conquered by Philip of Macedonia. This was followed by the conquests of Alexander the Great who was one of the greatest military commanders in history. He defeated the Persian Empire, and extended his empire to include most of the known world. His conquests extended as far east as Afghanistan and India, and as far west as Egypt. This led to the Hellenistic age, where Greek language and culture spread throughout the world. The word Hellenistic derives from "Ελλην" which means Greek. The city of Alexandra was founded by Alexander, and it became a major centre of learning. Alexander the Great received tuition from the philosopher Aristotle. However, Alexander's reign was very short as he died at the young age of 33 in 323 BC.

Early Greek mathematics commenced approximately 500–600 BC with work done by Pythagoras and Thales. Pythagoras was a sixth century philosopher and mathematician who had spent time in Egypt becoming familiar with Egyptian mathematics. He lived on the island of Samoa and formed a sect known as the Pythagoreans. This Pythagoreans were a secret society and included men and women. They believed in the transmigration of souls and believed that number was the fundamental building block for all things. They discovered the mathematics for harmony in music by discovering that the relationship between musical notes could be expressed in numerical ratios of small whole numbers. Pythagoras is credited with the discovery of Pythagoras's Theorem, although this theorem was probably known by the Babylonians about 1000 years before Pythagoras. The Pythagorean society was dealt a major blow[13] by the discovery of the incommensurability of the square root of 2: i.e., there are no numbers p, q such that $\sqrt{2} = p/q$. This dealt a major blow to their philosophy that number is the nature of being.

Thales was a sixth century (BC) philosopher from Miletus in Asia Minor who made contributions to philosophy, geometry and astronomy. His contributions to philosophy are mainly in the area of metaphysics, and he was concerned with questions on the nature of the world. His objective was to give a natural or scientific explanation of the cosmos, rather than relying on the traditional supernatural explanation of creation in Greek mythology. He believed that there was single substance that was the underlying constituent of the world, and he believed that this substance was water. It can only be speculated why he believed water to be the underlying substance but some reasons may be that water is essential to life; when a solid is compressed it is generally transformed to a liquid substance, and so on. Thales also contributed to mathematics [AnL:95] and there is a well-known theorem in Euclidean geometry named after him. It states that if A, B and C are points on a circle, and where the line AC is a diameter of the circle, then the angle $<ABC$ is a right angle.

---

[13] The Pythagoreans were a secret society and its members took a vow of silence with respect to this discovery. However, one member of the society is said to have shared the secret result with others outside the sect, and the apocryphal account is that he was thrown into a lake for his betrayal and drowned. They obviously took Mathematics seriously back then.

Euclid lived in Alexandria during the early Hellenistic period. He is considered the father of geometry in that he set out a systematic treatment of geometry starting from 5 axioms, 5 postulates and 23 definitions to derive and prove a comprehensive set of theorems. He is therefore the father of the axiomatic method for mathematics. His systematic account was published in the thirteen books of the Elements [Hea:56], and this has been used as a mathematics textbook for over 2000 years. It includes the treatment of geometry and number theory. His method of proof was generally constructive in that as well as demonstrating the truth of a theorem the proof would often include the construction of the required entity. However, he was also familiar with indirect proof as the argument to show that there are an infinite number of primes demonstrates:

1. Suppose there is a finite number of primes (say n primes).
2. Multiply all n primes together and add 1 to form N.

   i. $(N = p_1 * p_2 * \ldots * p_n + 1)$

3. N is not divisible by $p_1, p_2, \ldots, p_n$ as dividing by any of these gives a remainder of one.
4. Therefore, N must either be prime or divisible by some other prime that was not included in the list.
5. Therefore, there must be at least $n + 1$ primes.
6. This is a contradiction as it was assumed that there was a finite number of primes n.
7. Therefore, the assumption that there is a finite number of primes is false.
8. Therefore, there is an infinite number of primes.

Euclidean geometry included the parallel postulate or Euclid's fifth postulate. This postulate generated interest as many mathematicians believed that it was unnecessary and could be proved as a theorem by using the other axioms and postulates. It states that:

Definition 1.1 (Parallel Postulate) If a line segment intersects two straight lines forming two interior angles on the same side that sum to less than two right angles, then the two lines, if extended indefinitely, meet on that side on which the angles sum to less than two right angles.

This postulate was later proved to be independent of the other postulates. In the nineteenth century other geometries were developed that rejected the fifth postulate as formulated by Euclid. These include the hyperbolic geometry discovered independently by Bolyai and Lobachevsky, and elliptic geometry as developed by Riemann. The standard model of Riemannian geometry is the sphere where lines are great circles. Non-Euclidean geometries became important in the early twentieth century with the work done by Albert Einstein in the Theory of Relativity.

Euclid and contemporary Hellenistic mathematicians aimed to provide constructive solutions to problems. That is, the proof of the existence was generally accompanied by an actual construction of the solution using an unmarked straightedge and compass. The material in the Euclid's Elements is presented logically, and it is a systematic development of geomerty starting from the small set of axioms, postulates and definitions, and it leads to theorems derived logically from the axioms and postulates. Euclid's deductive method has influenced later mathematicians and scientists. There are some jumps in reasoning in The Elements, and Hilbert added extra axioms to Euclidean geometry to make it more complete in the late nineteenth century.

The Elements contains many well-known mathematical results such as:

- Pythagoras's Theorem
- Thales Theorem
- Sum of Angles in a Triangle
- Prime Numbers
- Greatest Common Divisor and Least Common Multiple
- Euclidean Algorithm
- Areas and Volumes
- Tangents to a point
- Algebra

The Euclidean algorithm is one of the oldest known algorithms and is employed to produce the greatest common divisor of two numbers. It is presented in the Elements but was known well before Euclid. The formulation of the gcd algorithm for two natural numbers a and b is as follows:

1. Check if b is zero. If so, then a is the gcd.
2. Otherwise, the gcd (a, b) is given by gcd $(b, a \bmod b)$.

It is also possible to determine integers p and q such that $ap + bq = \gcd(a, b)$.

The proof of the Euclidean algorithm is as follows. Suppose a and b are two positive numbers whose gcd has to be determined, and let r be the remainder when a is divided by b.

1. Clearly $a = qb + r$ where q is the quotient of the division.
2. Any common divisor of a and b is also a divider or r (since $r = a - qb$).
3. Similarly, any common divisor of b and r will also divide a.
4. Therefore, the greatest common divisor of a and b is the same as the greatest common divisor of b and r.
5. The number r is smaller than b and we will reach $r = 0$ in finitely many steps.
6. The process continues until $r = 0$.

Comment 1.1 Algorithms are fundamental in computing as they define the procedure by which a problem is solved. A computer program implements the algorithm in some programming language.

Eratosthenes was a Hellenistic mathematician and scientist who worked as librarian in the famous library in Alexandria. He devised a system of latitude and longitude, and became the first person to estimate of the size of the circumference of the earth. His approach to the calculation was as follows (Fig. 1.6):

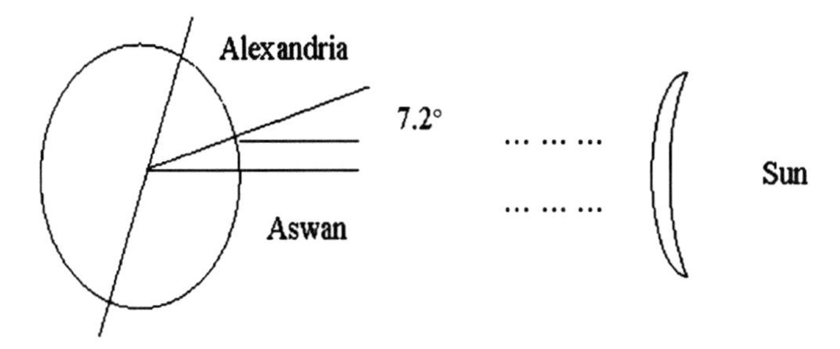

Fig. 1.6 Eratosthenes measurement of the circumference of the earth

1. On the summer solstice at noon in the town of Aswan[14] on the Tropic of Cancer in Egypt the sun appears directly overhead.
2. Eratosthenes believed that the earth was a sphere.
3. He assumed that rays of light came from the sun in parallel beams and reached the earth at the same time.
4. At the same time in Alexandria he had measured that the sun would be 7.2° south of the zenith.
5. He assumed that Alexandria was directly north of Aswan.
6. He concluded that the distance from Alexandria to Aswan was 7.2/360 of the circumference of the earth.
7. Distance between Alexandria and Aswan was 5000 stadia (approximately 800 km).
8. He established a value of 252,000 stadia or approximately 39,6000 km.

Eratosthene's calculation was within 1% of the true value of 40,008 km and was an impressive result for 200 BC. The errors in his calculation were due to:

1. Aswan is not exactly on the Tropic of Cancer but it is actually 55 km north of it.
2. Alexandria is not exactly north of Aswan and there is a difference of 3° longitude.
3. The distance between Aswan and Alexandtia is 729 km not 800 km.
4. Angles in antiquity could not be measured with absolute precision.
5. The angular distance is actually 7.08° and not 7.2°.

Eratosthenes also calculated the approximate distance to the moon and sun and he also produced maps of the known world. He developed a very useful algorithm for determining all of the prime numbers up to a specified integer. The method is known as the Sieve of Eratosthenes and the steps are as follows:

1. Write a list of the numbers from 2 to the largest number that you wish to test for primality. This first list is called A.

---

[14] The town of Aswan is famous today for the Aswan high dam which was built in the 1960s. There was an older Aswan dam built by the British in the late nineteenth century. The new dam led to a rise in the water level of Lake Nasser and flooding of archaeological sites along the Nile. Several sites such as Abu Simbel and the island of Philae were relocated to higher ground.

2. A second list B is created to list the primes. It is initially empty.
3. The number 2 is the first prime number and is added to the list of primes in B.
4. Strike off (or remove) all multiples of 2 from List A.
5. The first remaining number in List A is a prime number and this prime number is added to List B.
6. Strike off (or remove) this number and all multiples of this number from List A.
7. Repeat steps 5 through 7 until no more numbers are left in List A.

Comment 1.2 The Sieve of Eratosthenes method is a well-known algorithm for determining prime numbers. Computing students often implement this algorithm as an early computer assignment.

Archimedes was a mathematician, astronomer and philosopher who lived in Syracuse. He is famous for his discovery of the law of buoyancy that is known as Archimedes's principle:

> The buoyancy force is equal to the weight of the displaced fluid.

Archimedes is believed to have discovered the principle while sitting in his bath He was so overwhelmed with his discovery that he rushed out onto the streets of Syracuse shouting "Eureka", but forgot to put on his clothes to announce the discovery.

The weight of the displaced liquid will be proportional to the volume of the displaced liquid. Therefore, if two objects have the same mass, the one with greater volume (or smaller density) has greater buoyancy. An object will float if its buoyancy force (i.e., the weight of liquid displaced) exceeds the downward force of gravity (i.e., its weight). If the object has exactly the same density as the liquid, then it will stay still, neither sinking nor floating upwards.

For example, a rock is generally a very dense material and will generally not displace its own weight. Therefore, a rock will sink to the bottom as the downward weight exceeds the buoyancy weight. However, it the weight of the object is less than the liquid it would displace then it floats at a level where it displaces the same weight of liquid as the weight of the object.

Archimedes also made good contributions to mathematics including a good approximation to π, contributions to the positional numbering system, geometric series, and to maths physics. He also solved several interesting problems: e.g., the calculation of the composition of cattle in the herd of the Sun god by solving a number of simultaneous Diophantine equations. The herd consisted of bulls and cows with one part of the herd consisting of white, second part black, third spotted and the fourth brown. Various constraints were then expressed in Diophantine equations and the problem was to determine the precise composition of the herd. Diophantine equations are named after Diophantus worked who worked on number theory in the third century.

Archimedes also worked on another interesting problem to determine the number of grains of sands in the known universe. He challenged the prevailing view that the number of grains of sand was too large to be counted, and in order to provide an upper bound he needed to develop a naming system for large numbers. The largest number in common use at the time was a myriad (100 million) and a myriad is

10,000. Archimedes' numbering system goes up to $8*10^{16}$ and he also developed the laws of exponents: i.e., $10^a 10^b = 10^{a+b}$. His calculation of the upper bound includes not only the grains of sand on each beach but on the earth filled with sand and the known universe filled with sand. His final estimate of the number of grains of sand in a filled universe is an upper bound of 10^{64} for the number of grains of sand in a filled universe.

Is is possible that he may have developed the odometer,[15] and this instrument could calculate the total distance travelled on a journey. An odometer is described by the Roman engineer Vitruvius around 25 BC. It employed a wheel with a diameter of 4 feet, and the wheel turned 400 times in every mile.[16] The device included gears and pebbles and a 400 tooth cogwheel that turned once every mile and caused one pebble to drop into a box. The total distance travelled was determined by counting the pebbles in the box.

Aristotle was born in Macedonia and became a student of Plato in Athens. Plato had founded a school (known as Plato's academy) in Athens in the fourth century BC, and this school remained open until 529 A.D. Aristotle became a famous philosopher in his own right and he founded his own school (known as the Lyceum) in Athens. He was also the teacher of Alexander the Great. Aristotle made contributions to physics, biology, logic, politics, ethics and metaphysics. The work of Plato and Aristotle provided the foundations for Western philosophy (Fig. 1.7).

Aristotle's starting point to the acquisition of knowledge was the senses. He believed that the senses were essential to acquire knowledge. This position is the opposite from Plato who argued that the senses deceive and should not be relied upon. Plato's writings are mainly in dialogues involving his former mentor Socrates.[17] Most of Aristotle's writings are in treatise form although he wrote some dialogues in his early career. Aquinus,[18] a thirteenth century Christian theologian and philosopher, was deeply influenced by Aristotle, and referred to him as the philosopher. Acquinus was an empiricist (i.e., he believed that all knowledge was gained by sense experience), and he used some of Aristotle's arguments to offer five proofs of the existence of God. These arguments included the Cosmological argument and the Design argument. The Cosmological argument used Aristotle's

---

[15] The origin of the word "odometer" is from the Greek words 'οδοζ (meaning journey) and μετρον meaning (measure).

[16] The figures given here are for the distance of one Roman mile. This is given by $\pi 2^2 * 400 = 12.56 * 400 = 5024$ (which is less than 5280 feet for a standard mile in the Imperial system).

[17] Socrates was a moral philosopher who deeply influenced Plato. His method of enquiry into philosophical problems and ethics was by questioning. Socrates himself maintained that he knew nothing (Socratic ignorance). However, from his questioning it became apparent that those who thought they were clever were not really that clever after all. His approach obviously would not have made him very popular with the citizens of Athens. Socrates had consulted the oracle at Delphi to find out who was the wisest of all men, and he was informed that there was no one wiser than him. Socrates was sentenced to death for allegedly corrupting the youth of Athens, and the sentence was carried out by Socrates being forced to take hemlock (a type of poison). The juice of the hemlock plant was prepared for Socrates to drink.

[18] Aquinus's (or St. Thomas's) most famous work is Sumna Theologicae.

Fig. 1.7 Plato and Aristotle

ideas on the scientific method and causation. Acquius argued that there was a first cause and he deduced that this first cause is God.

1. Every effect has a cause
2. Nothing can cause itself
3. A causal chain cannot be of infinite length
4. Therefore there must be a first cause

Aristotle made important contributions to formal reasoning with his development of syllogistic logic and foundational work in modal logic. His collected works on logic is called the Organon and it was used in his school in Athens. Syllogistic logic (also known as term logic) consists of reasoning with two premises and one conclusion. Each premise consists of two terms and there is a common middle term. The conclusion links the two unrelated terms from the premises. This is best illustrated by an example:

Premise 1	All Greeks are Mortal
Premise 2	Socrates is a Greek.
	....
Conclusion	Socrates is Mortal

In this example the common middle term is "Greek", and this term appears in the two premises. The two unrelated terms from the premises are "Socrates" and "Mortal". The relationship between the terms in the first premise is that of the universal: i.e., anything or any person that is a Greek is mortal. The relationship between the terms in the second premise is that of the particular: i.e., Socrates is a person that is a Greek. The conclusion from the two premises is that Socrates is mortal: i.e., a particular relationship between the two unrelated terms "Socrates" and "Mortal".

The example above is an example of a valid syllogistic argument. Aristotle studied the various possible syllogistic arguments and determined those that were valid and those that were invalid. There are several candidate relationships that may potentially exist between the terms in a premise. These include (Table 1.1):

Table 1.1 Syllogisms: Relationship between terms

Relationship	Abbr.
Universal Affirmation	A
Universal Negation	E
Particular Affirmation	I
Particular Negation	O

In general, a syllogistic argument will be of the form:

S x M
M y P
.....
S z P

where x, y, z may be universal affirmation, universal negation, particular affirmation and particular negation. Syllogistic logic is described in more detail in [ORg:06]. Aristotle's work was highly regarded in classical and medieval times, and was believed to be a fully worked out system. Kant believed that there was nothing else to invent in Logic after the work of Aristotle. There was another competing system of logic proposed by the Stoics in Hellentistic times: i.e., an early form of propositional logic that was developed by Chrysippus[19] in the third century BC. Later work in the nineteenth century by George Boole led to propositional logic, and later work by Frege and others led to predicate calculus. Aristotelian logic is mainly of historical interest today.

The Greeks invented a number of mechanical devices to assist with problem solving, and one of the most famous of these was the Antikythera [Pri:59]. This was an ancient mechanical device designed to calculate astronomical positions. An ancient Antikythera was discovered in 1902 in a week off the Greek island of Antikythera, and dates from about 80 BC. It is one of the oldest known geared devices, and it

[19] Chrysippus was the head of the Stoics in the third century BC.

is believed that it was used for calculating the position of the sun, moon, stars and planets for a particular date entered. The device is comparable in the complexity of its parts and gear arrangement as clocks in the eighteenth century.

The Romans appear to have been aware of a device similar to the Antikythera, as a device that is capable of calculating the position of the planets is mentioned by Cicero. The island of Antikythera was well-known in the Greek and Roman period for its displays of mechanical engineering. A model of how the Antikythera might have worked is available, and according to that model, the front dial shows the annual progress of the Sun and Moon through the zodiac against the Egyptian calendar, with the other rear dials providing specialised information. It is debatable as to how accurate the model is with respect to the actual device. Other models proposed include that the device acted as a planetarium.

1.5 The Romans

Rome is said to have been founded[20] by Romulus and Remus about 750 BC. Early Rome covered a small part of Italy but it gradually expanded in size and importance. Rome destroyed Carthage[21] in 146 BC to become the major power in the Mediterranean. Julius Caesar (Gaius Iulius Caesar) initially conquered the Gauls in 58 BC (Fig. 1.8).

The Gauls consisted of several Celtic[22] tribes who were disunited. Vercingetorix was the leader of the Arverni tribe and he succeeded briefly in uniting the Celts. However, Caesar finally defeated Vercingetorix at the siege of Alesia in 52 BC. Roman merchants needed to develop accounting systems to track their trade across the Roman Empire. The Hellenistic world was colonized by the Romans,[23] and the Romans became familiar with Greek culture and mathematics.

The Romans introduced their own number system where Roman letters represented numbers (Fig. 1.9):

---

[20] The Aenid by Virgil suggests that the Romans were descended from survivors of the Trojan war, and that Aeneas brought surviving Trojans to Rome after the fall of Troy.

[21] Carthage was located in Tunisia, and the wars between Rome and Carthage are known as the Punic wars. Hannibal was one of the great Carthaginan military commanders, and during the second Punic war, he brought his army to Spain, marched through Spain and crossed the Pyrenees. He then marched along southern France and crossed the Alps into Northern Italy. His army also consisted of war elephants. Rome finally defeated Carthage and levelled the city.

[22] The Celtic period commenced around 1000 BC in Hallstaat (near Salzburg in Austria). The Celts were skilled in working with Iron and Bronze, and they gradually expanded into Europe. They eventually reached Britain and Ireland around 600 BC. The early Celtic period was known as the "Hallstaat period" and the later Celtic period is known as "La Téne". The later La Téne period is characterised by the quality of ornamentation produced. The Celtic museum in Hallein in Austria provides valuable information and artefacts on the Celtic period. The Celtic language would have similarities to the Irish language. However, the Celts did not employ writing, and the Ogham writing developed in Ireland was employed later in the early Christian period.

[23] The Romans did not make any major advances on Hellenistic Mathematics.

Fig. 1.8 Julius Caesar

Fig. 1.9 Roman numbers

I = 1
V = 5
X = 10
L = 50
C = 100
D = 500
M = 1000

A Roman number consists of a sequence of Roman letters and there were rules employed in the evaluation. The rules specified that if a number follows a smaller number then the smaller number is subtracted from the large: e.g., IX represents 9 and XL represents 40. Similarly, if a smaller number followed a larger number they were generally added: e.g., MCC represents 1200. They had no zero in their system. Roman numerals are still used today in page numbering for books or on the faces of clocks.

Calculations with Roman numerals was cumbersome, especially operations that involved multiplication or division. In practice, an abacus was often employed to perform the calculation. An abacus consists of several columns in which pebbles are placed. Each column represented powers of 10: i.e., 10^0, 10^1, 10^2, 10^3, etc. The column to the far right represents one; the column to the left 10; next column to the left 100; and so on. Pebbles (calculi) were placed in the columns to represent different numbers: e.g., the number represented by an abacus with 4 pebbles on the far right; 2 pebbles in the column to the left; and 3 pebbles in the next column to the left is 324. Calculations were performed by moving pebbles from column to column. The operator of the abacus needed to be properly trained to be effective.

The Roman merchant needed to perform calculations to keep track of trade within the Roman Empire. They introduced a set of weights and measures (including the *libra* for weights and the *pes* for lengths). The merchants also developed an early banking system to provide loans for business. They commenced minting money about 290 BC. The Romans also made contributions to calendars and the Julian calendar was introduced in 45 BC by Julius Caesar. It has a regular year of 365 days divided into 12 months and a leap day is added to February every four years. It remained in use up to the twentieth century, but has since been replaced by the Gregorian calendar. The problem with the Julian calendar is that too many leap years are added over time. The Gregorian calendar was first introduced in 1582.

The Romans employed the available mathematics that had been developed by the Greeks. Caesar's Cipher was employed by Caesar on his military campaigns in order to safely communicate important messages to his generals. It is one of the simplest and widely known encryption techniques, and involves the substitution of each letter in the plaintext (i.e., the original message) by a letter a fixed number of positions down in the alphabet. For example, a shift of 3 positions cause the letter B to be replaced by E, the letter C by F, and so on. The Caesar cipher is easily broken, as the frequency distribution of letters may be employed to determine the mapping. However, given that Caesar was essentially dealing with Gaulish tribes who were mainly illiterate, and who certainly lacked knowledge of cryptology and frequency distribution of the letters in the alphabet, it is likely to have provided good security. The translation of the Roman letters by the Caesar cipher (with a shift key of 3) can be seen by the following table. It shows each letter of the alphabet and the corresponding cipher symbol that it is mapped on to (Table 1.2):

Table 1.2 Caesar cipher

Alphabet Symbol	abcde	fghij	klmno	pqrst	uvwxyz
Cipher Symbol	dfegh	ijklm	nopqr	stuvw	xyzabc

The process of enciphering a message (i.e., plaintext) simply involves looking up each letter in the plaintext and writing down the corresponding cipher letter. For example, the enciphering of the plaintext message "summer solstice" involves the following:

Plaintext	Summer Solstice
Cipher Text	vxpphu vrovwleh

The process of deciphering a cipher message involves doing the reverse operation: i.e., for each cipher letter the corresponding plaintext letter is identified from the table.

Cipher Text	vxpphu vrovwleh
Plaintext	Summer Solstice

The encryption can also be represented using modular arithmetic by first using the numbers 0–25 to represent the alphabet letters, and then using addition (modula 26) to perform the encryption. That is, the encoding of the plaintext letter represented by the number x is given by:

$$c = x + 3 \ (\text{mod } 26)$$

Similarly, the decoding of a cipher letter represented by the number c is given by:

$$x = c - 3 \ (\text{mod } 26)$$

The emperor Augustus[24] employed a similar substitution cipher (with a shift key of 1). The Caesar cipher was still in use up to the early twentieth century. However, by then frequency analysis techniques were available to break the cipher. The Vignère cipher uses a Caesar cipher with a different shift at each position in the text. The value of the shift to be employed with each plaintext letter is defined using a repeating keyword.

The famous library in Alexandria was once the largest library in the world. It was build during the Hellenistic period in the third century BC. Caesar's campaign in Egypt in 48 BC caused damage to the library, and the library was finally destroyed by fire in 391 A.D. The new library in Alexandria was inaugurated in 2003 on the site of the old library.

1.6 Islamic Influence

Islamic mathematics refers to mathematics developed in the Islamic world from the birth of Islam in the early seventh century up until the seventeenth century. The Islamic world commenced with Mohammed in Mecca, and spread throughout the Middle East, North Africa and Spain. Islamic scholars translated the works of

[24] Augustus was the first Roman emperor and his reign ushered in a period of peace and stability following the bitter civil wars. He was the adopted son of Julius Caesar and was called Octavion before he became emperor. The earlier civil wars were between Caesar and Pompey, and following Caesar's assassination civil war broke out between Mark Anthony and Octavion. Octavion defeated Anthony and Cleopatra at the battle of Actium.

the Greeks into Arabic, and this led to the preservation of the Greek texts during the Dark ages in Europe. Further, the Islamic scholars developed the existing mathematics further. The Islamic contribution filled the void that followed the end of the Roman Empire in the sixth century A.D. The Moors[25] invaded Spain in the eighth century A.D., and they ruled large parts of the Iberian Peninsula for several centuries. The Moorish influence[26] in Spain continued until the time of the Catholic Monarchs[27] in the fifteenth century. Ferdinand and Isabella united Spain, defeated the Moors, and expelled them from Spain.

The Islamic mathematicians and scholars were based in several countries including Iran, Iraq, Turkey, North Africa and Spain. Early work commenced in Baghdad, and the mathematicians were influenced by the work of Hindu mathematicians who had introduced the decimal system and decimal numerals. There was a renaissance in European learning and interest in mathematics in the seventeenth century, and the Islamic texts played a key part in the revival.

Many caliphs (Muslim rulers) were enlightened and encouraged scholarship in mathematics and science. The initial work done was the translation of the existing Greek texts, and this led to a centre of translation and research in Baghdad. The translations were done as part of the research effort and included the works of Euclid, Archimedes, Apollonius and Diophantus. Al-Khwarizmi[28] made contributions to early classical algebra, and the word algebra comes from the Arabic word "*al jabr*" that appears in a text book by Al-Khwarizmi.

Early work in algebra had been done by the Babylonians, Egyptians and Greeks. The Babylonians had a general procedure for solving quadratic equations but there was limited use of symbols for unknowns. The Greeks represented quantities as geometrical magnitudes. Later Greek mathematicians such as Diophantus developed algebra for solving Diophantine equations. This included solving equations in several unknowns. The Islamic contribution to algebra was an advance on the achievements of the Greeks. They developed a broader theory that treated rational and irrational numbers as algebraic objects, and moved away from the Greek concept of mathematics as being essentially Geometry. Later Islamic scholars built on Al-Khwarizmi's work, and applied algebra to arithmetic and geometry. This included contributions to reduce geometric problems such as duplicating the cube to algebraic problems. Eventually this led to the use of symbols in the fifteenth century such as:

$$x^n \cdot x^m = x^{m+n}.$$

[25] The origin of the word "Moor" is from the Greek work μυορоζ meaning very dark. It referred to the fact that many of the original Moors who came to Spain were from Egypt, Tunisia and other parts of North Africa.

[26] The Moorish influence includes the construction of various castles (*alcazar*), fortresses (*alcalzaba*) and mosques. One of the most striking Islamic sites in Spain is the palace of Alhambra in Granada, and this site represents the zenith of Islamic art.

[27] The Catholic Monarchs refer to Ferdinand of Aragon and Isabella of Castille who married in 1469. They captured Granada (the last remaining part of Spain controlled by the Moors) in 1492.

[28] The origin of the word algorithm is from the name of the Islamic scholar Al-Khwarizmi.

The poet Omar Khayyam was also a mathematician. He did work on the classification of cubic equations with geometric solutions. Others also applied algebra to geometry, and aimed to study curves by using equations. Other scholars made contributions to the theory of numbers: e.g., a theorem that allows pairs of amicable numbers to be found. Amicable numbers are two numbers such that each is the sum of the proper divisors of the other. They were aware of Wilson's theory in number theory: i.e., for p prime then p divides $(p-1)!+1$. This result was proved formally by Lagrange in the eighteenth century.

Moorish Spain became a centre of learning, and this led to Islamic and other scholars coming to study in the universities in Spain. This includes scholars such as Averros and Avicenna who provided commentaries on the work of Aristotle. Many texts on Islamic mathematics were translated from Arabic into Latin, and this helped to start the renaissance in learning and mathematics in Europe.

1.7 Chinese and Indian Mathematics

The development of mathematics in China was independent of developments in other countries. This was due to the geographical position of China, and its ability to absorb other cultures into its own without changing its own. The development of mathematics in China commenced about 1000 BC. The Chinese approach to mathematics differed from the Greeks, in that its focus was on problem solving rather than on conducting formal proofs. Hellenistic mathematics employed an axiomatic approach with axioms and rules of deduction. Chinese mathematics was pragmatic, and was concerned with finding the solution to practical problems such as the calendar, the prediction of the positions of the heavenly bodies, land measurement, conducting trade, and the calculation of taxes.

The Chinese employed counting boards as mechanical aids for calculation from the fourth century BC. Counting-boards are similar to abaci and are usually made of wood or metal, and contained carved grooves between which beads, pebbles or metal discs were moved. The abacus is a device, usually of wood having a frame that holds rods with freely-sliding beads mounted on them. It is used as a tool to assist calculation, and it is useful for keeping track of the sums, the carrys, and so on of calculations.

Early Chinese mathematics was written on bamboo strips and included work on arithmetic and astronomy. The Chinese method of learning and calculation in mathematics was learning by analogy. This involves a person acquiring knowledge from observation of how a problem is solved, and then applying this knowledge for problem-solving to similar kinds of problems.

The Chinese had their version of Pythagoras's Theorem and applied it to practical problems. One well-knowm Chinese mathematical book is the Mathematical Treatise in Nine Sections. This dates from the thirteenth century and it was used as a textbook for several hundred years. It included the Chinese remainder theorem, the formula for finding the area of a triangle, as well as showing how polynomial equations (up to degree ten) could be solved. Other Chinese mathematicians showed how geometric problems could be solved by algebra, how roots of polynomials could be solved, how quadratic and simultaneous equations could be solved, and how the area of various geometric shapes such as rectangles, trapezia and circles could be computed. Chinese mathematicians were familiar with the formula to calculate the volume of a sphere. The best approximation that the Chinese had of π was 3.14159, and this was obtained by Hui by approximations from inscribing regular polygons with 3×2^n sides in a circle. Hui seems to have been familiar with the idea of a limit, as his approximation to π is achieved using an iterative approach with each iteration achieving a closer approximation to π.

The Chinese made contributions to number theory including the summation of arithmetic series and solving simultaneous congruences. The Chinese remainder theorem deals with finding the solutions to a set of simultaneous congruences in modular arithmetic. Chinese astronomers made accurate observations which were used to produce a new calendar in the sixth century. This was known as the Taming Calendar and it was based on a cycle of 391 years.

Indian mathematicians have made major contributions to the development of mathematics. One key contribution of Indian mathematicians is to the development of the decimal notation for numbers that is now used throughout the world. The decimal system was developed in India sometime between 400 BC and 400 AD. Indian mathematicians also invented zero and negative numbers, and also did early work on the trigonometric functions of sine and cosine The knowledge of the decimal numerals reached Europe through Arabic mathematicians, and the resulting system is known as the Hindu-Arabic numeral system.

The Sulva Sutras is a Hindu text that documents Indian mathematics and it dates from about 400 BC. The Indian mathematicians were familiar with the statement and proof of Pythagoras's theorem, and were familiar with Rational numbers, quadratic equations, as well as the calculation of the square root of 2 to five decimal places.

Panini was a fifth century BC. Indian mathematician and linguist who did pioneering work on phonetics and morphology for the Sanskrit language. His work on grammar allowed sentences to be formed from a set of rules, and is the earliest known work on linguistics and formal grammars.

1.8 Review Questions

1. Describe the number systems employed by the various civilizations discussed in this chapter and discuss the strengths and weaknesses of each system.
2. Describe ciphers used during the Roman civilization and write a program to implement one of these. What were the disadvantages of these ciphers and how would you improve upon them?

3. Discuss the nature of an algorithm and its importance in computing. Describe any algorithm that you are familiar and implement in a programming language of your choice.
4. Discuss the working of an abacus and its application to calculation.
5. Discuss syllogistic logic and identify strengths and weaknesses of the logic. Discuss the similarities and differences between syllogistic logic and propositional and predicate logic.

1.9 Summary

The last decades of the twentieth century have witnessed a proliferation of high-tech computers, mobile phones, and information technology. Software is now pervasive and it is included in automobiles, airplanes, televisions, and mobile communication. It is only in recent decades that technology has become an integral part of the western world, and the pace of change has been extraordinary. It has led to increases in industrial productivity and potentially allows humans the freedom to engage in more creative and rewarding tasks.

This chapter considered the contributions of early civilisations in providing a primitive foundation for what has become computer science. It included a discussion on the Babylonians, the Egyptians, the Greeks and the Romans as well as contributions from Islamic scholars.

The Babylonian civilization flourished from about 2000 BC and they produced clay cuneiform tablets containing mathematical texts. These included tables for multiplication, division, squares, and square roots as well as the calculation of area and the solution of linear and quadratic equations.

The Egyptian Civilization developed along the Nile from about 4000 BC and lasted until the Roman Empire. They used mathematics for practical problem solving, and this included measuring time, measuring the annual Nile flooding, calculating the area of land, solving baking problems. The use of mathematics was essential in constructing pyramids.

The Greeks made major contributions to western civilization including mathematics, logic, philosophy, politics, drama, and architecture. Euclid developed a systematic treatment of geometry starting from a small set of axioms, postulates and definitions to derive and prove a comprehensive set of theorems. Euclid's Elements has been in use as a textbook for over 2000 years.

The Romans developed a cumbersome number system that is still used in clocks today. They also developed the Julian calendar and employed simple ciphers to ensure that information communicated was kept confidential. The Islamic contribution helped to preserve the earlier work of the Greeks, and they also developed mathematics and algebra further. Later Islamic scholars applied algebra to arithmetic and geometry.

Chapter 2
Foundations

Key Topics

Boolean Algebra
Difference Engine
Analytic Engine
Information Theory
Cryptography
Turing Machines
Artificial Intelligence
The Enigma Machine
Von Neumann Architecture
The ENIAC
The Manchester Mark 1

2.1 Introduction

This chapter considers various pioneers in the history of computing. These include figures such as Boole, Babbage, Turing, Shannon and Von Neumann. George Boole was a nineteenth century English mathematician who is considered one of the fathers of computing. His calculus of logic is known as Boolean Algebra and is the foundation of all modern computers.

Charles Babbage was a nineteenth century scientist who did pioneering work on the Difference Engine (a sophisticated calculator that could be used for the production of mathematical tables). Babbage also designed the Analytic Engine, and this was the design of the world's first mechanical computer. It included a processor, memory, and a way to input information and output results. However, the machine was never built during Babbage's lifetime.

Babbage intended that the program be stored on read-only memory using punch cards and that input and output for the Analytic Engine be carried out using punch cards. He intended that the machine would be able to store numbers and intermediate results in memory where they could then be processed. He even intended that the machine would be capable of parallel processing, where several calculations could be performed at once.

Turing was a famous twentieth century English mathematician and computer scientist. His mathematical Turing Machine proved that anything that is computable is computable by this theoretical machine. He also made contributions to the British war effort during the second-world war while working at Bletchey Park. This team at Bletchey succeeded in breaking the German Enigma codes. Turing also did important work in Artificial Intelligence and he devised the famous "Turing Test" as a test of machine intelligence. He also did work at Manchester University prior to his premature death in the early 1950s.

Claude Shannon was an American mathematician and engineer who made fundamental contributions to computing. He was the first person to see the applicability of Boolean algebra to simplify the design of circuits and telephone routing switches. His influential Masters Thesis is a key milestone in computing, and it shows how to lay out circuits according to Boolean principles. It provides the theoretical foundation of switching circuits, and his insight of using the properties of electrical switches to do Boolean logic is the basic concept that underlies all electronic digital computers.

Shannon's later work at Bell Labs laid the foundation of modern information theory. Information theory is concerned with the problem of reliable transfer of messages from a source point to a destination point. This includes the transfer of messages over any communications medium: for example, television, radio, telephone, and computers. The fundamental problem of communication is to reproduce at a destination point either exactly or approximately a message that has been sent from a source point. The problem is that information may be distorted by noise and the received message may differ from the original sent. Shannon provided a mathematical definition and framework for information theory that remains the standard today. He also made a significant contribution to the emerging field of cryptography.

John von Neumann was a Hungarian mathematician who made fundamental contributions to mathematics, set theory, computer science, and the American atomic bomb programme. His PhD thesis showed how the paradoxes in set theory could be avoided by employing the concept of a class.

He gave his name to the well-known von Neumann architecture used in almost all computers. The key features of von Neumann architecture are an arithmetic unit for performing basic arithmetic operations; a control unit; an input-output unit and memory. The control unit executes the instructions stored in memory, and uses a program counter to fetch the next instruction in memory, which it then executes. The input-output unit allows the computer to interact with the outside world, and the one-dimensional memory tcontains the program instructions and data.

Von Neumann also created the field of cellular automata, and invented the merge-sort algorithm (in which the first and second halves of an array are each sorted recursively and then merged). There is an annual IEEE von Neumann medal that is awarded by the IEEE to researchers for outstanding achievements in computer science.

2.2 Boole

Boole (1815–1864) was born of humble parents in Lincoln, England. His father was a cobbler with an active mind who was interested in mathematics and optical instruments. He provided an early influence on his son by teaching him mathematics, and teaching him how to make optical instruments. George Boole (Fig. 2.1) inherited his father's interest in knowledge, and was self-taught in mathematics. He was taught Latin by a tutor, but was self-taught in Greek. He taught in various schools near Lincoln, and developed his mathematical knowledge by working his way through Newton's Principia, as well as applying himself to the work of mathematicians such as Laplace and Lagrange.

He published regular papers from his early twenties onwards, and these included contributions to probability theory, differential equations, and finite differences. He is especially remembered for his major contribution to Boolean algebra which is the foundation for modern computing. Boole is therefore considered (along with Babbage) to be one of the fathers of computing. Boole, never actually built any computer, as he lived well before the computer age. However, the mathematical foundation provided by Boole was exactly what was required for telephone switching and computing, and it was Claude Shannon who saw the potential of Boole's work, and who brought it to practical fruition. Boole's work remains important today.

Fig. 2.1 George Boole

Boole was interested in formulating a calculus of reasoning, and in 1847 he published a pamphlet titled "Mathematical Analysis of Logic" [Boo:48]. This article developed novel ideas on a logical method, and he argued that logic should be considered as a separate branch of mathematics, rather than being considered a part of philosophy. Boole argued that there are mathematical laws to express the operation of reasoning in the human mind, and he showed how Aristotle's syllogistic logic could be rendered as algebraic equations. This publication was well received, and the British mathematician and logician De-Morgan[1] spoke highly of it. There was regular correspondence between Boole and De-Morgan on logic. Boole was interested in obtaining a university position, but due to his lack of a formal university education it was therefore difficult for him to achieve this goal. However, the value of his publications were recognized in Britain, and in view of the excellence of his work,[2] and especially the pamphlet on Logic, he was awarded the position as the first professor of mathematics at the newly founded Queens University of Cork,[3] Ireland in 1849.

Boole's influential paper on a calculus of logic introduced two quantities 0 and 1. He used the quantity 1 to represent the universe of thinkable objects (i.e., the universal set), with the quantity 0 representing the absence of any objects (i.e., the empty set). He then employed symbols such as x, y, z, etc., to represent collections or classes of objects given by the meaning attached to adjectives and nouns. Next, he introduced three operators ($+$, $-$, and $\times$) that combined classes of objects. For example, the expression xy (i.e., x multiplied by y) combines the two classes x, y to form the new class xy (i.e., the class whose objects satisfy the two meanings represented by the classes x and y). Similarly, the expression $x+y$ combines the two classes x, y to form the new class $x+y$ (that satisfies either the meaning represented by class x or class y). The expression $x - y$ combines the two classes x, y to form the new class $x - y$. This represents the class (that satisfies the meaning represented by class x but not class y). The expression $(1 - x)$ represents objects that do not have the attribute that represents class x.

Thus, if x = black and y = sheep, then xy represents the class of black sheep. Similarly, $(1 - x)$ would represents the class obtained by the operation of selecting all things in the world except black things; $x(1 - y)$ represents the class of all things that areblack but not sheep; and $(1 - x) \cdot (1 - y)$ would give us all things that are

[1] De-Morgan was a 19th British mathematician based at University College London, and is well-known for De-Morgan's laws in Set Theory and Logic: $\neg$ (A $\cup$ B) = $\neg$A $\cap$ $\neg$B and $\neg$ (A $\vee$ B) = $\neg$A $\wedge$ $\neg$B. In Boole's notation this would be expressed as $(1 - (A + B)) = (1 - A) \cdot (1 - B)$. This is written as $\overline{(A + B)} = \bar{A}\ \bar{B}$ in Shannon's notation.

[2] Boole was awarded the Royal Medal from the Royal Society of London in 1844 in recognition of his publications. The Irish Mathematician, Sir. Rowan Hamilton, (who discovered Quaternions) was another famous recipient of this prize. Hamilton was awarded the medal in 1835. Boole was elected a fellow of the Royal Society in 1857.

[3] Queens University Cork is now called University College Cork (UCC) and has about 15,000 students. It is located in Cork city in the south of Ireland. Queens University Belfast and Queens University Galway (now NUI, Galway) were founded the same year.

neither sheep nor black. Similarly, if z = goats then $y + z$ represents the class that are either sheep or goats. He showed that these symbols obeyed a rich collection of algebraic laws and could be added, multiplied, etc., in a manner that is similar to real numbers. Boole showed how these symbols could be used to reduce propositions to equations, and algebraic rules could be used to solve the equations. The algebraic rules satisfied by his system included:

1.	$x + 0 = x$	(Additive Identity)
2.	$x + (y + z) = (x + y) + z$	(Associativity)
3.	$x + y = y + x$	(Commutativity)
4.	$x + (1 – x) = 1$	(Not operator)
5.	$x1 = x$	(Multiplicative Identity)
6.	$x0 = 0$	
7.	$x + 1 = 1$	
8.	$xy = yx$	(Commutativity)
9.	$x(yz) = (xy)z$	(Associativity)
10.	$x(y + z) = xy + xz$	(Distributive)
11.	$x(y – z) = xy – xz$	(Distributive)
12.	$x^2 = x$	(Idempotent)
13.	$x^n = x$	

These operations are similar to the modern laws of set theory with the set union operation represented by "+", and this is the "or" operation in modern Boolean algebra. The set intersection operation is represented by multiplication, and this is the "and" operation in Boolean Algebra. The universal set is represented by "1" is denoted by U in modern set theory, and the empty set "0" is denoted by ∅. The laws of associativity and distribution hold. Finally, the set complement operation is given by $(1 - x)$, and this is the "not" operation of Boolean Algebra.

Boole applied the symbols to encode propositions of Aristotle's Syllogistic Logic, and he showed how the propositions could be reduced to the form of equations. These equations allowed conclusions to be derived from premises by eliminating the middle term in the syllogism. Syllogistic logic had been in use for over 2000 years, and was the main form of logic employed in the nineteenth century. It remains of historical interest today as it has largely been replaced by predicate logic.

Boole refined his ideas further in his book "An Investigation of the Laws of Thought" [Boo:58] published in 1854. This book and his earlier paper on logic contain the concepts which have come to be known collectively as Boolean algebra. His book aimed to identify the fundamental laws underlying reasoning in the human mind, and to give expression to these laws in the symbolic language of a calculus. Boole considered the equation $x^2 = x$ to be one of the fundamental laws of though. It allows the principle of contradiction (i.e., that it is impossible for a being to possess an attribute and at the same time not to possess it) to be derived from this fundamental law of thought.

$$
\begin{aligned}
& x^2 = x \\
& \Rightarrow x - x^2 = 0 \\
& \Rightarrow x(1 - x) = 0
\end{aligned}
$$

For example, if x represents the class of horses then $(1 - x)$ represents the class of "not-horses". The product of two classes represents a class whose members are common to both classes. Hence, $x(1 - x)$ represents the class whose members are at once both horses and "not-horses", and the equation $x(1 - x) = 0$ expresses that fact that there is no such class whose members are both horses and "not-horses". That is, there are no members of such a class and it is the empty set.

Boole made contributions to other areas apart from logic. He published papers on differential equations and on finite differences.[4] He also published a book on differential equations which was used as a text-book at Cambridge University in England. He also contributed to the development of probability theory.

Boole married Mary Everest in 1855 and they lived in Lichfield Cottage in Ballintemple, Cork. She was a niece of the surveyor of India, Sir George Everest, after whom the world's highest mountain[5] in the Himalayas is named. The Booles had five daughters, and one of their daughters was the author of the "The Gadfly".[6] Boole died prematurely (at the early age of 49) from pneumonia in 1864. He got soaked while walking two miles in the rain from his home in Ballintemple to teach his class at Queens University, Cork. He is buried in the graveyard of St. Michaels Church in Blackrock, Cork.

Queens University Cork honoured his memory by installing a stained glass window in the *Aula Maxima* of the college. This shows Boole writing at a table with Aristotle and Plato in the background. University College Cork is the modern name for Queens University Cork, and it named the new library after Boole in 1983. The Mathematics department at University College Cork holds an annual competition that is open to students of mathematics or engineering, and the winner is awarded the annual Boole prize. The prize recognises the recipient as having potential in mathematics. Further information on George Boole including his work and working relationship with Queens University Cork is in [Bar:69, McH:85].

Boole left a major legacy to the world as the design of all modern binary digital computers is dependent on Boolean algebra. The Boolean logical operations are implemented by electronic AND, OR and NOT gates, and from these fundamental building blocks more complex circuits may be designed for operations such as arithmetic. In view of this legacy, Boole is rightly considered to be one of the founding fathers of computing.

---

[4] Finite Differences are a numerical method to solve differential equations.

[5] Everest is known as Chomolungma ("Mother of the Universe") to the Tibetans.

[6] This is a novel about the struggle of an international revolutionary. Shostakovich wrote the score for the film of the same name that appeared in 1955.

2.2.1 Boolean Algebra

Boolean algebra was developed by George Boole in the nineteenth century. This section presents Boolean algebra in the more modern notation of the propositional calculus that computer science students are familiar with. Boolean algebra consists of propositions that are either true or false. For example, the proposition "$2+2 = 4$" is true, whereas the proposition "$2 * 5 = 11$" is false. Variables (e.g., A, B, etc.) are used to stand for propositions, and propositions may be combined using logical connectives to form new propositions. The standard logical connectives are "and", "or" and "not", and these are represented by the symbols "$\wedge$", "$\vee$" and "$\neg$" respectively. There are other logical connectives that may be used such as implication ($\Rightarrow$) and equivalence ($\Leftrightarrow$). These connectives may be expressed using the other logical connectives (e.g., $A \Rightarrow B$ is equivalent to $\neg A \vee B$). Propositions must be either true or false.

There are several well-known properties of Boolean algebra such as the commutative, associative, and distributive properties. These are summarized in Table 2.1 below:

Table 2.1 Properties of Boolean algebra

Property	Example
Commutative	$A \wedge B = B \wedge A$
	$A \vee B = B \vee A$
Associative	$A \wedge (B \wedge C) = (A \wedge B) \wedge C$
	$A \vee (B \vee C) = (A \vee B) \vee C$
Identity	$A \wedge \text{True} = \text{True} \wedge A = A$
	$A \vee \text{False} = \text{False} \vee A = A$
Distributive	$A \wedge (B \vee C) = (A \wedge B) \vee (A \wedge C)$
	$A \vee (B \wedge C) = (A \vee B) \wedge (A \vee C)$
De-Morgan	$\neg (A \wedge B) = \neg A \vee \neg B$
	$\neg (A \vee B) = \neg A \wedge \neg B$
Idempotent	$A \wedge A = A$
	$A \vee A = A$

The commutative property expresses the fact that the order of the two operands may be reversed without affecting the truth value of the resulting proposition. The associative property states that the conjunction or disjunction operators are associative. This means that for a Boolean expression that consists of conjunctions or disjunctions only, that the order of execution is not relevant. In other words, the same result is obtained by applying the logical operator to the first two operands and then the applying it to the third, as applying it to the second and third and then applying the result to the first operand.

The Boolean constant "True" is the identity operation for conjunction. In other words, the conjunction of any operand A with the Boolean value "True" yields the proposition A. Similarly, the Boolean constant "False" is the identity operation for disjunction. Conjunction distributes over disjunction and vice versa. De Morgan's Law states that the negation of the conjunction of two operands is the same as the disjunction of the negation of each operand. The idempotent property states that the application of the conjunction or disjunction logical operator to two operands that are A yields the proposition A.

Truth Tables enable the truth-values of a compound proposition to be determined from its underlying propositions. The conjunction of A and B $(A \wedge B)$ is true if and only if both A and B is true. The disjunction of A and B $(A \vee B)$ is true if either A or B is true. These are as defined in Table 2.2

Table 2.2 Truth tables for conjunction and disjunction

A	B	A ∧ B	A	B	A ∨ B
T	T	T	T	T	T
T	F	F	T	F	T
F	T	F	F	T	T
F	F	F	F	F	F

Disjunction is also known as the "inclusive or" operation. There are other logical connectives such as implication, equivalence, "exclusive or", and so on. For a more detailed explanation see [ORg:06]. The "not" operator (¬) is a unary operator such that ¬ A is true if A is false, and is false if A is true. It is given by the following truth table (Table 2.3).

Table 2.3 Truth table for Not operation

A	¬ A
T	F
F	T

Boolean algebra is used extensively by programmers in conditional and iterative statements. Complex Boolean expressions may be formed from simple Boolean expressions using the logical connectives ("AND", "OR" and "NOT"). For example, $x > 2$, $x < 6$ are two simple Boolean expressions. A more complex Boolean expression that yields the set $\{3,4,5\}$ is given by:

$$x > 2 \text{ AND } x < 6$$

Boolean algebra plays an essential role in sorting and searching. For example, the following example in SQL selects data from the employee table where the age of the employee is greater than 20.

```
SELECT name, age, salary
FROM Employees
WHERE age > 20
```

2.2.2 Foundations of Computing

Boole's work on Boolean Algebra remained relatively unknown for many years as it seemed to have little practical use to society. However, Claude Shannon's research at Massachusetts Institute of Technology in the 1930s showed that Boolean Algebra could be employed in telephone routing switches. He showed that it could play a role in the design of systems used in telephone routing and that conversely that these circuits could solve Boolean algebra problems.

Modern electronic computers use the properties of electrical switches to do logic, with high voltage representing the Boolean state "on" and low voltage representing the Boolean state "off". Boolean algebra is basic to the design of digital computer circuits, and all modern computers use Boolean Logic. Computer microchips contain thousands of tiny electronic switches arranged into logical gates. The basic logic gates are AND, OR and NOT. These gates may be combined in various ways to allow the computer to perform more complex tasks. Each gate has binary value inputs and outputs.

The output from the gate is a single high or low voltage logical conclusion. The voltage represents the binary "true" or "false" or equivalently "on" or "off" or the binary value "1" or "0". The example in Fig. 2.2 below is that of an "AND" gate and this gate produces the binary value "1" as output only if both inputs are "1". That is, the AND gate will produce a TRUE result only if all inputs are TRUE. Otherwise, the result will be FALSE (i.e., binary 0).

Figure 2.3 is that of an "OR" gate and this gate produces the binary value "1" as output if any of its inputs is "1". That is, the OR gate will produce a TRUE result if any of its input is TRUE otherwise it will produce the output FALSE.

Finally, a NOT gate accepts only a single input, either TRUE or FALSE, which it reverses (Fig. 2.4). That is, if the input is "1" the value "0" is returned and vice versa. The NOT gate may be combined with the AND or the OR gate to yield the NAND gate (NOT AND) or the NOR gate (NOT OR).

The power of the AND, OR and NOT gates may be seen in the example of a half adder in Fig. 2.5. It shows how these gates may be combined to do elementary binary arithmetic. The addition of 1 + 0 is considered. The inputs to the top OR gate are 1 and 0 and this yields the result of 1. The inputs to the bottom AND gate are 1 and 0 which yields the result 0, which is then inverted through the NOT gate to yield

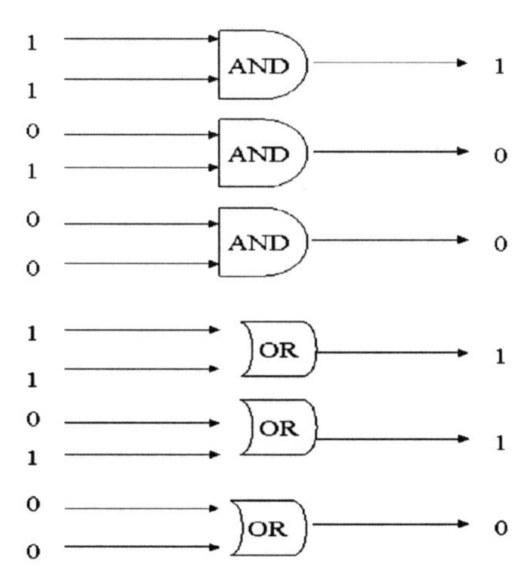

Fig. 2.2 Binary AND operation

Fig. 2.3 Binary OR operation

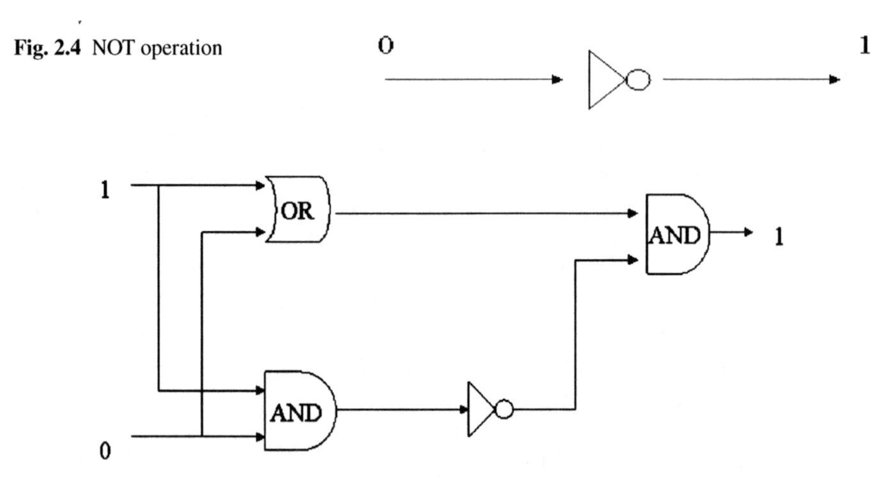

Fig. 2.4 NOT operation

Fig. 2.5 Half-adder
The half-adder may be extended to a full adder that provides a carry for addition.

binary 1. Finally, the last AND gate receives two 1's as input and returns a binary 1 result. Thus, the result of 1 + 0 is 1. This example of the half-adder correctly computes the addition of two arbitrary binary digits. However, it does not compute a carry for the addition of 1 + 1.

2.3 Babbage

Charles Babbage (Fig. 2.6) is considered (along with George Boole) to be one of the fathers of computing. He was born in Devonshire, England in 1791 and was the son of a banker. He studied Mathematics at Cambridge University in England and was appointed to the Lucasian Chair in Mathematics at Cambridge in 1828. He was a profound thinker and make contributions in several areas including mathematics, statistics, astronomy, philosophy, railways and lighthouses. He founded the British Statistical society and the Royal Astronomical Society. His pioneering work in computing remains well-known today.

Babbage was interested in mathematical tables but recognised that there was a high error rate in the tables due to the human error introduced during their calculation. He was interested in solving this problem, and his approach was to investigate a method by which the calculation of tables could be done mechanically. This would thereby eliminate the errors introduced by humans in their computation. There had been some experimental work done on calculating machines by Pascal and Leibnitz, and Babbage became interested in designing a calculating machine from the early 1820s.

He designed the Difference Engine (No. 1) in 1821 and this was a special purpose machine for the production of mathematical tables. A difference engine is essentially a mechanical calculator (analogous to modern electronic calculators), and it

Fig. 2.6 Charles Babbage

was designed to compute polynomial functions. It could also compute logarithmic and trigonometric functions such as sine or cosine, as these may be approximated by polynomials.[7] The accurate approximation of trigometric, exponential and logarithmic functions by polynomials depends on the degree of the polynomials, the number of decimal digits that it is being approximated to, and the error function. A higher degree polynomial is able to approximate the function more accurately than a lower degree polynomial. Accurate trigometric tables were essential for navigation, and the Difference Engine offered the potential for accurate computation by polynomial approximations. Babbage produced prototypes for parts of the Difference Engine but he never actually completed the engine. The Difference Engine envisaged by

[7] The power series expansion of the Sine function is given by $\sin(x) = x - x^3/3! + x^5/5! - x^7/7! + \ldots$ The power series expansion for the Cosine function is given by $\cos(x) = 1 - x^2/2! + x^4/4! - x^6/6! + \ldots$ Functions may be approximated by interpolation and the approximation of a function by a polynomial of degree n requires $n + 1$ points on the curve for the interpolation. That is, the curve formed by the polynomial of degree n that passes through the $n + 1$ points on the function to be approximated is an approximation to the function.

Babbage was intended to operate on 6th order polynomials of 20 digits. A polynomial of degree 6 is of the form $p(x) = ax^6 + bx^5 + cx^4 + dx^3 + ex^2 + fx + g$.

2.3.1 Difference Engine

The first working difference engine was built in 1853 by the Swedish engineers George and Edward Schuetz. The Schuetz brothers based their plans on Babbage's design, and received funding from the Swedish government to build the machine. The Schuetz machine could compute polynomials of degree 4 on 15-digit numbers. A copy of the 3rd Scheutz Difference Engine is on display in the Science Museum in London.

This machine was the first to compute and print mathematical tables mechanically. Today, the computation of mathematical tables is done by electronic computers. However, in the nineteenth century this task was performed manually and it was error prone. The working engine produced by the Schuetz brothers was accurate, and it was used by scientists and engineers to perform their calculations. Babbage's Difference Engine showed the potential of mechanical machines as a tool for scientists and engineers. Today, the availability of modern electronic calculators means that students no longer need to concern themselves with the production of tables.

The difference engine consists of a number of columns (1 to N). Each column is able to store one decimal number, and the numbers are represented by wheels. The Difference Engine (No. 1) has seven columns with each column containing 20 wheels. Each wheel consists of ten teeth and these represent the decimal digits. Each column could therefore represent a decimal number with up to 20 digits. The seven columns allowed the representation of polynomials of degree six. The engine could store only seven decimal numbers at any one time, and a Difference Engine with N columns may store N numbers at any one time. Surprisingly, this is sufficient for the engine to do its work to calculate the required tables.

The only operation that the Difference Engine can perform is the addition of the value of column n + 1 to column n, and this results in a new value for column n. Column N can only store a constant and column 1 displays the value of the calculation for the current iteration. The machine needs to be programmed prior to execution, and this is done by setting initial values to each of the columns. Column 1 is set to the value of the polynomial at the start of computation; column 2 is set to a value derived from the first and higher derivatives of the polynomial for the same value of x. Each of the columns from 3 to N is set to a value derived from the n−1st and higher derivatives of the polynomial.

The Schuetz's difference engine was an implementation of Babbage's design, and it is comprised of shafts, shelves and wheels. The scientist could set numbers on the wheels,[8] and turn a crank to start the computation. Then, by reading down each shaft, he could find the result of a calculation. The difference engine was able

[8] Each wheel represented a decimal digit and each wheel consisted of 10 teeth to perform the decimal representation.

to print out the answers to the computation. The decimal numbering system was employed, and there was also a carry mechanism.

The difference engine is unable to perform multiplication or division, and is therefore unable to calculate the value of a polynomial directly. Instead, once the initial value of the polynomial and its derivatives are calculated for some value of x, the difference engine can calculate any number of nearby values using the method of finite differences. This method is used by numerical analysts, and the objective is to replace computational intensive tasks that involve multiplication or division by an equivalent computation that just involves addition or subtraction which is less computationally intensive.

2.3.2 Finite Differences

A finite difference[9] is a mathematical expression of the form $f(x+h) - f(x)$. If a finite difference is divided by h, then the resulting expression is similar to a differential quotient, except that it is discrete.

$$\frac{\Delta f}{h} = \frac{f(x+h) - f(x)}{h} \approx f'(x) \text{ (when } h \text{ is small)}$$

Finite differences may be applied to approximate derivatives and they are often used to find numerical solution to differential equations. The finite difference may be considered to be an operator Δ that takes a function f as an argument and returns Δf. By employing the Taylor series expansion it can be seen that $\Delta = e^{hD} - 1$.[10] Finite difference approximation to higher order derivatives may also be obtained. For example:

$$f''(x) \approx \frac{f'(x+h) - f'(x)}{h} \approx \frac{f(x+2h) - 2f(x+h) + f(x)}{h^2} = \frac{\Delta^2 f}{h^2}$$

That is,

$$f^n(x) \approx \frac{\Delta^n f}{h^n} \quad \text{(when } h \text{ is small)}$$

The method of finite differences plays a key role in the production of tables for the polynomial using the Difference Engine. Consider the quadratic polynomial $p(x) = 2x^2 + 3x + 1$ and consider Table 2.4.

---

[9] Finite differences may be further differentiated into forward finite differences (Δf), backward differences (∇f) and central finite differences δf.

[10] D represents the derivative operator that maps a function f to its derivative f'. D^n represents the operator that maps a function to its nth derivative.

Table 2.4 Finite differences

x	$f(x)$	Diff. 1	Diff. 2
1	6		
2	15	9	
3	28	13	4
4	45	17	4
5	66	21	4

The 1st difference is computed by subtracting two adjacent entries in the column of $f(x)$. For example, $15 - 6 = 9$; $28 - 15 = 13$; and so on. Similarly, the 2nd difference is given by subtracting two adjacent entries in the Difference first column: e.g., $13 - 9 = 4$; $17 - 13 = 4$; and do on. The entries in the 2nd difference column are the constant 4. In fact, for any n-degree polynomial the entries in the n-difference column is always a constant.

The Difference Engine performs the computation of the table in a similar manner, although the approach is essentially the reverse of the above. The key point to realise is that once you have the first row of the table, that the rest of the table may then be computed using just additions of pairs of cells in the table. The first row of Table 2.5 is given by the cells 6, 9 and 4, and this allows the rest of the table to be determined.

The numbers in the table below have been derived by simple calculations from the first row. The procedure for calculation of the table is as follows.

1. The Difference 2 column is the constant 4.
2. The calculation of the cell in row i for the Difference 1 column is given by Diff. $1(i\text{–}1)$ + Diff. 2 $(i\text{–}1)$
3. The calculation of the cell in row i for the function column is given by $f(i\text{–}1)$ + Diff. $1(i\text{–}1)$

In other words, to calculate the value of a particular cell, all that is required is to add the value in the cell immediately above it to the value of the cell immediately to its right. Therefore, in the second row, the calculations 6 + 9 yields 15, and 9 + 4 yields 13, and, since the last column is always a constant it is just repeated. Therefore, the second row is 15, 13 and 4 and f (2) is 15. Similarly, the third row yields 15 + 13 = 28, 13 + 4 = 17, and so on. This is the underlying procedure of the Difference Engine.

Table 2.5 Finite differences

x	$f(x)$	Diff. 1	Diff. 2
1	6	9	4
2	15	13	4
3	28	17	4
4	45	21	4
5	66	25	4

The initial problem is then to compute the first row as this enables successive rows to be computed. The Difference Engine cannot compute a particular value, and its approach is to calculate successive values starting from a known one and the first row of differences. The computation of the first row for the example above is easy, but it is more difficult for more complex polynomials. The other problem is to find a suitable polynomial to represent the function that you wish to calculate. This may be done by interpolation. However, once these problems are solved the engine produces pages and columns full of data.

Babbage had received funding to build the Difference Engine but never delivered a complete system. He received over £17K of taxpayers funds, but for various reasons he only produced prototypes of the intended machine. Therefore, the return to the exchequer was limited, and the government obviously had little enthusiasm for the other projects that Babbage was proposing. The project to build the Difference Engine was officially cancelled by the British government in 1842.

Various prototypes of parts of the Difference Engine machine had been built by Babbage and his engineer Joseph Clement, but these machines were limited to the computation of quadratic polynomials of six digit numbers. The Difference Engine envisaged by Babbage was intended to operate on 6th order polynomials of 20 digits. Such a machine would take up the space of a complete room.

Babbage designed an improved difference engine (No. 2) in 1849 (Fig. 2.7). It could operate on 7th order differences (i.e., polynomials of order 7) and 31-digit numbers. A polynomial of degree 7 is of the form $p(x) = ax^7 + bx^6 + cx^5 + dx^4 + ex^3 + fx^2 + gx + h$. The machine consisted of eight columns with each column

Fig. 2.7 Difference engine No. 2
Photo courtesy of Wikipedia.

consisting of thirty one wheels.[11] However, no one was interested in his design and it was never built in his lifetime. It was built 150 years later (in 1991) by the science museum in London to mark the two hundredth anniversary of Babbage's birth in. The science museum also built the printer that Babbage had designed. Both the difference engine and the printer worked correctly according to Babbage's design (after a little debugging).

2.3.3 Analytic Engine

A key weakness of the Difference Engine is that it requires the intervention of humans to perform the calculation. Babbage recognised this weakness of the engine, and he proposed a revolutionary idea to resolve this. His plan was to construct a new machine that would be capable of analysis, and this machine would be capable of executing all possible tasks that may be expressed in algebraic notation. The new machine would do substantially more than the arithmetical calculations performed by the difference engine, which was essentially limited to the production of tables. The Analytic Engine envisioned by Babbage consisted of two parts (Table 2.6):

Babbage saw the operation of the Analytic Engine as analogous to the operation of the Jacquard loom.[12] The latter is capable of weaving any design which may be conceived by man. A weaving company employs a team of skilled artists to design patterns for the loom, and the pattern is then executed by the loom. The pattern is designed by punching holes on a set of cards and the cards are ordered so that each card represents successive rows in the design. The cards are then placed in the Jacquard loom, and the exact pattern as designed by the artist is produced by the loom.

Table 2.6 Analytic engine

Part	Function
Store	This contains the variables to be operated upon as well as all those quantities which have arisen from the result of intermediate operations.
Mill	The mill is essentially the processor of the machine into which the quantities about to be operated upon are brought.

[11] The thirty one wheels allowed 31-digit numbers to be represented as each wheel could represent a decimal digit.

[12] The Jacquard loom was invented by Joseph Jacquard in 1801. It is a mechanical loom which used the holes in punch cards to control the weaving of patterns in a fabric. The use of punched cards allowed complex designs to be woven from the pattern defined on the punched card. Each punched card corresponds to one row of the design and the cards were appropriately ordered. It was very easy to change the pattern of the fabric being weaved on the loom, as this simply involved changing cards.

The Jacquard loom was the first machine to use punch cards to control a sequence of operations. The machine did not perform computation, but its ability to change the pattern of what was being weaved by changing cards gave Babbage the inspiration to use punched cards to store programs to perform the analysis and computation in the Analytic Engine.

The use of the punched cards in the design of the Analytic Engine was extremely powerful, as it allowed the various formulae to be manipulated in a manner dictated by the programmer. The cards commanded the analytic engine to perform various operations and to return a result. Babbage distinguished between two types of punched cards:

- Operation Cards
- Variable Cards

Operation cards are used to define the operations to be performed, whereas the variable cards define the variables or data that the operations are performed upon. This planned use of punched cards to store programs in the Analytic Engine is similar to the idea of a stored computer program in Von Neumann architecture. However, Babbage's idea of using punched cards to represent machine instructions was over 100 years before Von Neumann's architecture. Babbage's Analytic Engine is therefore an important step in the history of computing.

The Analytic Engine was designed by Babbage in 1834 [Bab:42], and was the design of the world's first mechanical computer. It included a processor, memory, and a way to input information and output results. However, the machine was never built during Babbage's lifetime due to difficulty in achieving funding from the British Government. The government was not convinced that further funding should be given to start work on another machine given that his first machine (i.e., the Difference Engine) had not been built.

Babbage intended that the program be stored on read-only memory using punch cards, and that the input and output would be carried out using punch cards. He intended that the machine would store numbers and intermediate results in memory that could then be processed. There would be several punch card readers in the machine for programs and data, and that the machine should be able to perform conditional jumps. Babbage planned for parallel processing where several calculations could be performed at once.

Lady Ada (Fig. 2.8) was introduced into Babbage's ideas on the analytic engine at a dinner party. She was a mathematician and the daughter of the poet Lord Byron. She was fascinated by the idea of the analytic engine and communicated regularly with Babbage on ideas on applications of the engine. She predicted that such a machine could be used to compose music, produce graphics, as well as solving mathematical and scientific problems. She suggested to Babbage that a plan be written for how the engine would calculate Bernoulli numbers. This plan is considered to be the first computer program, and Lady Ada Lovelace is therefore considered the first computer programmer. The Ada programming language developed in the United States is named in her honour.

Fig. 2.8 Lady Ada Lovelace

Babbage died in 1871 and his brain was preserved in alcohol. His brain was then dissected over 30 years later in 1908 and is on view today in the science museum in London.

2.4 Formalism

Gottlob Frege was a nineteenth century German mathematician and logician. He invented a formal system which is the basis of modern predicate logic. It included axioms, definitions, universal and existential quantification, and formalization of proof. This formal system would essentially become the predicate calculus. His objective was to show that mathematics was reducible to logic but this project failed as one of the axioms that he had added to his system proved to be inconsistent.

This inconsistency was pointed out by Bertrand Russell, and is known as Russell's paradox.[13]

The expressions in a formal system are all terms, and a term may be simple or complex. A simple term may be an object such as a number, and a complex term may be an arithmetic expression such as $4^3 + 1$. The complex term is formed via functions, and for the expression above the function employed is the cube function with argument 4.

The sentences of the logical system are complex terms that denote the truth-values of true or false. The sentences may include the binary function for equality (such as $x = y$), and this returns true if x is the same as y, and false otherwise. Similarly, more complex expression such as $f(x, y, z) = w$ is true if $f(x, y, z)$ is identical with w, and false otherwise. Frege represented statements such as "5 is a prime" by "$P(5)$" where $P()$ is termed a concept. The statement $P(x)$ returns true if x is prime. His approach was to represent a predicate as a function of one variable which returns a Boolean value of true or false.

Formalism was proposed by Hilbert (Fig. 2.9) as a foundation for mathematics in the early twentieth century. The motivation for the programme was to provide a secure foundations for mathematics and to resolve the contradictions in set theory identified by Russell's paradox.

A formal system consists of a formal language, a set of axioms and rules of inference. It is generally intended to represent some aspect of the real world. Hilbert's programme was concerned with the formalization of mathematics (i.e., the axiomatization of mathematics) together with a proof that the axiomatization was consistent. The specific objectives of Hilbert's programme were to:

- Develop a formal system where the truth or falsity of any mathematical statement may be determined.
- A proof that the system is consistent (i.e., that no contradictions may be derived).

A proof in a formal system consists of a sequence of formulae, where each formula is either an axiom or derived from one or more preceding formulae in the sequence by one of the rules of inference.

Hilbert outlined twenty-three key problems that needed to be solved by mathematicians, and he believed that every problem could be solved (i.e., the truth or falsity of any mathematical proposition could be determined). Further, he believed that formalism would allow a mechanical procedure (or algorithm) to determine whether a particular statement was true or false. This problem is known as the decision problem (Entscheidungsproblem).The question had been considered by Leibnitz in the seventeenth century. He had constructed a mechanical calculating machine, and wondered if a machine could be built that could determine whether particular mathematical statements are true or false.

Definition 2.1 (Algorithm) An algorithm (or procedure) is a finite set of unambiguous instructions to perform a specific task.

[13] Russell's paradox is discussed in section 2.7.

Fig. 2.9 David Hilbert

The term “algorithm” is named after the Persian mathematician Al-Khwarizmi as discussed in Chapter 1. The concept of an algorithm was defined formally by Church in 1936 with the lambda calculus, and independently by Turing with his theoretical Turing machines. These two approaches are equivalent.

Hilbert believed that every mathematical problem could be solved and therefore expected that the formal system of mathematics would be decidable:

Definition2.2 (Decidability) Mathematics is decideable if the truth or falsity of any mathematical proposition may be determined by an algorithm.

Church and Turing independently showed this to be impossible in 1936. Turing showed that decidability was related to the halting problem for Turing machines, and therefore if first-order logic were decidable then the halting problem for Turing machines could be solved. However, he had already proved that there was no general algorithm to determine whether a given Turing machine halts. Therefore, first order logic is undecidable.

A well-formed formula is valid if it follows from the axioms of first-order logic. A formula is valid if and only if it is true in every interpretation of the formula in the model. Gödel proved that first order predicate calculus is complete. i.e., all truths in the predicate calculus can be proved in the language of the calculus.

Definition 2.3 (Completeness) A formal system is complete if all the truths in the system can be derived from the axioms and rules of inference.

Gödel showed in 1930 that fist order arithmetic was incomplete: i.e., there are truths in first-order arithmetic that cannot be proved in the language of the axiomatization of first-order arithmetic.

There are first-order theories that are decideable. However, first-order logic that includes Peano's axioms of arithmetic cannot be decided by an algorithm.

2.5 Turing

Alan Turing (Fig. 2.10) was a British mathematician who made fundamental contributions to mathematics and computer science. In his short but highly productive life he made contributions to computability, cryptology, the breaking of the Enigma naval codes, theoretical work on Turing machines, practical work in the development of the first stored program computer in England, and contributions to the emerging field of Artificial Intelligence with the Turing Test.

He was born in London 1912 and his father worked as a magistrate and tax collector in the Indian civil service. He attended the Sherbone[14] which was a famous public school in England. The school was very strong in the classics rather than in science and mathematics. Turing's interests were in science, mathematics and chess and he was weak in writing and literature. Turing applied himself to solving advanced problems in science and mathematics, and this included studying Einstein's work on relativity.

Turing excelled at long-distance running at the Sherbone, and in later life he completed the marathon in 2 hours and 46 minutes. This time would have made him a candidate for the 1948 Olympic games that were held in Wembley stadium in London. However, unfortunately he was injured shortly before the games, and this prevented him from future competition. Turing had few close friends at school with the exception of Christopher Morcom who was a gifted student. They worked closely together on science experiments and mathematics. Turing Morcom died of Tuberculosis three years later.

Turing attended Kings College, Cambridge from 1931 to 1934, and he graduated with a distinguished degree. He was elected a fellow of Kings College in 1935, and the computer room at King's college is named after him. He published a key paper on computability in 1936. This paper introduced a theoretical machine known as the Turing machine, and this mathematical machine is capable of performing

[14] The Sherbone is located in Dorset, England. Its origins go back to the 8th century when the school was linked with the Benedictine Abbey in the town.

Fig. 2.10 Alan Turing

any conceivable mathematical problem that has an algorithm. The Church-Turing thesis that states that anything that may be computed may be computed by a Turing machine. There are other equivalent definitions of computability such as Church's lambda calculus.

2.5.1 Turing Machines

A Turing machine consists of a head and a potentially infinite tape (Fig. 2.11) that is divided into frames. Each frame on the tape may be either blank or printed with a symbol from a finite alphabet of symbols. The input tape may initially be blank or have a finite number of frames containing symbols. At any step, the head can read the contents of a frame. The head may erase a symbol on the tape, leave it unchanged, or replace it with another symbol. It may then move one position to the right, one position to the left, or not at all. If the frame is blank, the head can either leave the frame blank or print one of the symbols.

Turing believed that every calculation could be done by a human with finite equipment and with an unlimited supply of paper to write on. The unlimited supply of paper is formalised in the Turing machine by a paper tape marked off in squares.

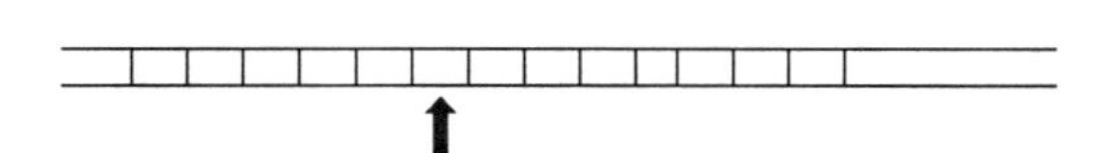

Fig. 2.11 Potentially infinite tape

The finite number of configurations of the Turing machine were intended to represent the finite states of mind of a human calculator.

Definition 2.4 (Turing Machine) A Turing machine M = (Q, Γ, b, Σ, δ, q_0, F) is a 7-tuple as defined formally in [HoU:79] where:

- Q is a finite set of *states.*
- Γ is a finite set of the *tape alphabet/symbols.*
- b ∈ Γ is the *blank symbol* (This is the only symbol that is allowed to occur infinitely often on the tape during each step of the computation).
- Σ is the set of input symbols and is a subset of Γ (it does not include the blank symbol b).
- $\delta : Q \times \Gamma \rightarrow Q \times \Gamma \times \{L, R\}$ is the transition function. This is a partial function called where L is left shift, R is right shift.
- $q_0 \in Q$ is the initial state.
- $F \subseteq Q$ is the set of final or accepting states.

The Turing machine is very simple machine, and it is equivalent to an actual physical computer in the sense that it can compute exactly the same set of functions. A Turing machine is much easier to analyse and prove things about than a real computer. However, Turing machines are not suitable for programming, and therefore do not provide a good basis for studying programming and programming languages.

A Turing machine is essentially a finite state machine (FSA) with an unbounded tape. The machine may read from and write to the tape. The FSA is essentially the control unit of the machine, and the tape is essentially the store. The tape is potentially infinite and unbounded, whereas real computers have a large but finite store. However, the store in a real computer may be extended with backing tapes and disks, and in a sense may be regarded as unbounded. However, the maximum amount of tape that may be read or written within n steps is n.

A Turing machine has an associated set of rules that defines its behaviour, and the actions that a machine may perform are defined by the transition function. A Turing machine may be programmed to solve any problem for which there is an algorithm. However, if the problem is unsolvable then the machine will either stop or compute forever. The solvability of a problem may not be determined beforehand. There is, of course, some answer (i.e., either the machine halts or it computes forever).

Turing showed that there was no solution to the Entscheidungsproblem problem (i.e., the decision problem) posed by David Hilbert. This was discussed earlier, and it asserted that the truth or falsity of a mathematical problem can always be determined, and by a mechanical procedure. Hilbert was asserting that first-order logic is decidable: i.e., there is a decision procedure to determine if a sentence in the formula is valid. Turing was skeptical on the decidability of first-order logic, and he believed that whereas some problems may be solved, some others cannot be solved. The Turing machine played a key role in refuting Hilbert's claim of the decidability of first-order logic, as whatever is computable may be computed on a Turing Machine.

The question as to whether a given Turing machine halts or not can be formulated as a first-order statement. If a general decision procedure exists for first-order logic, then the statement of whether a given Turing machine halts or not is within the scope of the decision algorithm. However, Turing had already proved that the halting problem for Turing machines is not computable: i.e., it is not possible algorithmically to decide whether a give Turing machine will halt or not. Therefore, there is no general algorithm that can decide whether a given Turing machine halts. In other words, there is no general decision procedure for first-order logic. The only way to determine whether a statement is true or false is to try to solve it.

Turing also introduced the concept of a Universal Turing Machine and this machine is able to simulate any other Turing machine. Turing's results on computability were proved independently of Church's lambda calculus equivalent results in computability.[15] Turing's studied in Priceton University in 1937 and 1938 and was awarded a PhD from the university in 1938. He was awarded his PhD at the age of 25, and his research supervisor was Alonzo Church.[16] His results on computability and decidability were proved when he was just 23. He returned to Cambridge in 1939 and attended lectures by Wittgenstein. However, Turing and Wittgenstein disagreed on formalism and on mathematics.

2.5.2 Enigma Codes

Turing and others contributed to Britain's defence and to the eventual defeat of Nazi Germany during the second world war. Turing's contribution was the role that he played at Bletchey Park in breaking the Enigma codes. These codes were used by the Germans for the transmission of naval messages. They allowed messages to be passed secretly to submarines using encryption, and this was designed to prevent third parties from intercepting the messages. The plaintext (i.e., the original message) was converted by the Enigma machine (Fig. 2.12) into the encrypted text, and these messages were then transmitted by the Germans to their submarines in the Atlantic or their bases throughout Europe.

These messages contained top-secret information on German submarine and naval activities in the Atlantic, and the threat that they posed to British and Allied shipping. The team at Bletchey Park succeeded in breaking of the Enigma codes,

[15] However, Turing machines are in my view more intuitive than the lambda calculus.

[16] Alonzo Church was a famous American mathematician and logician who developed the lambda calculus. He also showed that Peano arithmetic and first order logic were undecidable. Lambda calculus is equivalent to Turing machines and whatever may be computed is computable by Lambda calculus or a Turing machine. This is known as the Church-Turing Thesis.

Fig. 2.12 The Enigma machine
Courtesy of Wikipedia.

and therefore saved many Allied lives in the Atlantic. This contributed to the Allied victory and the shortening of the war.

Turing worked in Hut 8 in the codebreaking centre at Bletchey Park which was located in north-west London. He designed an electromechanical machine known as the bombe (Fig. 2.13), and this machine was designed to help break the Enigma codes by finding the right settings for the Enigma machine. This became the main tool to read the Enigma traffic during the war. It was first installed in early 1940, and by the end of the war over 200 bombes were in operation. The bombe weighed over a ton and it was named after a cryptological device designed in 1938 by the Polish cryptologist Marian Rejewski.

A standard Enigma machine employed a set of three rotors. Each rotor could be in any of 26 positions. The bombe tried each possible rotor position and applied a test. The test eliminated almost all the 17,576 positions (i.e., 26^3) and left a smaller number of cases to be dealt with. The test required the cryptologist to

Fig. 2.13 Cardboard replica of bombe
Courtesy of Wikipedia.

have a suitable “crib”: i.e., a section of ciphertext for which he could guess the corresponding plaintext.

For each possible setting of the rotors, the bombe employed the crib to perform a chain of logical deductions. The bombe detected when a contradiction had occurred; it then ruled out that setting and moved onto the next. Most of the possible settings would lead to contradictions and could then be discarded. This would leave only a few settings to be investigated in detail. Turing and the team at Bletchey Park made fundamental contributions to breaking the German Enigma machine.

Turing was engaged briefly to Joan Clarke who was a co-worker at Bletchey Park. However, their engagement ended by mutual agreement as Turing was a homosexual. He travelled to the United States in late 1942 and worked with US cryptanalysts on bombe construction. He also worked with Bell Labs to assist with the development of secure speech. He returned to Bletchley Park in early 1943 and designed a portable machine for secure voice communications. It was developed too late to be used in the war. Turing was awarded the Order of the British Empire[17] (OBE) in 1945 in recognition of his wartime services to Britain.

Turing worked as a scientific officer at the National Physical Laboratory (NPL) after the war, and he worked on the design of the Automatic Computing Engine (ACE) (Fig. 2.14).

This was the first complete design of a stored-program computer in Britain. The ENIAC had been developed in the United States, and this was a huge machine that did numerical calculations. The ACE computer was intended to be a smaller and more advanced computer than ENIAC. However, there were delays with funding to start the project and Turing became disillusioned with NPL. He moved to the Uni-

[17] Of course, the British Empire was essentially over at the end of the second-world war with the United States becoming the major power in the world both economically and militarily. India achieved independence in 1947 and former colonies gradually declared independence from Britain in the coming years. The Irish Free State was awarded dominion status in 1921 after the Irish war of independence. It remained neutral during the second-world war and became a republic in 1949.

Fig. 2.14 NPL Pilot ACE
Courtesy of NPL © Crown Copyright 1950.

versity on Manchester to work on the software for the Manchester Mark 1 computer. He also did some work on a chess program for a computer that did not yet exist.

2.5.3 Turing Test in AI

Turing contributed to the debate concerning artificial intelligence in his 1950 paper on Computing, machinary and intelligence [Tur:50]. Turing's paper considered whether it could be possible for a machine to be conscious and to think. He predicted that it would be possible to speak of "thinking machines", and he devised a famous experiment that would allow a computer to be judged as a conscious and thinking machine. This is known as the "Turing Test". The test itself is an adaptation of a well-known party game which involves three participants. One of them, the judge, is placed in a separate room from the other two: one is a male and the other is a female. Questions and responses are typed and passed under the door. The objective of the game is for the judge to determine which participant is male and which is female. The male is allowed to deceive the judge whereas the female is supposed to assist.

Turing adapted this game by allowing the role of the male to be played by a computer. If the judge could not tell which of the two participants was human or machine, then the computer could be considered intelligent. That is, if a computer can convince a human that it is also a human, then the machine must also be considered to be intelligent and to have passed the "Turing Test". The test is applied to test the linguistic capability of the machine rather than to the audio capability, and the conversation between the judge and the parties is conducted over a text only channel.

Turing's work on "thinking machines" caused a great deal of public controversy, as defenders of traditional values attacked the idea of machine intelligence. Turing's paper led to a debate concerning the nature of intelligence. There has been no machine developed, to date, that has passed the Turing test, and there is unlikely to be one in the near future.

Turing strongly believed that machines would eventually be developed that would stand a good chance of passing the "Turing Test". He considered the operation of "thought" to be equivalent to the operation of a discrete state machine. Such a machine may be simulated by a program that runs on a single, universal machine, i.e. a computer. He did consider various objections to his position on artificial intelligence, and attempted to refute these objections in his 1950 paper. These include (Table 2.7):

Table 2.7 Turing and AI

Argument	Description
Theological Objections	God has given an immortal soul to every man and woman. Thinking is a function of man's immortal soul. God has not given an immortal soul to animals or inanimate machines. Therefore, animals or machines cannot think.
Head in the Sand objections	The consequences of thinking machines would be dreadful to mankind. Let us hope that this never happens as it would undermine the superiority of man over animals and machines.
Mathematical Objections	There is a limitation to the power of discrete machines, and consequently there are certain things that these machines cannot do. Goedel's Theorem is one example of a limitation of what machines may do. However, there are also limits on what humans may do.
Consciousness	The argument is that a machine cannot feel or experience emotion such as joy, anger, happiness, and so on. Turing argued that this position is similar to the difficulty in knowing that other humans can experience emotions.
Lady Lovelace's objection	This comes from a statement made by Lady Lovelace that the Analytic Engine can do only whatever we know or tell it to do. In other words, a machine can never do anything really new. However, as Turing points out, machines and computers often surprise us with new results.
Informality of Behaviour	This argument is along the lines that if each man has a set of laws (or rules of behaviour) that regulate his life then he would be no better than a machine. But there are no such rules and therefore men cannot be machines. However, the statement that there are no such rules or laws of behaviour is debatable.
Learning Machines	If a machine is to possess human like intelligence it will need to be capable of learning. That is, the machine will need to be a learning machine that acquires knowledge by experience as well as having an initial core set of knowledge inserted into the machine. Turing argued that researchers should aim to produce a computer with a child like mind and develop the knowledge of the computer by teaching it.

There are a number of recent objections to the Turing test, including Searle's Chinese room argument, and this essentially argues that the simulation of human language behaviour is weaker than true intelligence. This is discussed in a later chapter.

Turing later did some work in biology in the area of morphology which is concerned with the form and structure of plants and animals.

2.5.4 Later Life

Turing was a homosexual and the Victorian laws in England still applied during this period. The British had prosecuted the famous Irish playright, Oscar Wilde, in 1895 under Section 11 of the 1885 Criminal Law Amendment Act. This legislation prohibited acts of gross indecency between males. Turing had a brief relationship with Arnold Murray in the early 1950s. However, Murray and an accomplice burgled Turing's house when the relationship ended in 1952. Turing reported the matter to the police, and at the criminal trial allegations of homosexuality were made against Turing.

He was charged under the 1895 act and was convicted. He was then given a choice between imprisonment or probation. The terms of the probation were severe, and required him to receive oestrogen hormone injections for a year. There were side affects with the treatment and it led to depression and despair. Further, his conviction led to a removal of his security clearance and the end of his work on cryptography for the British government.

Turing committed suicide in June 1954. The medical examiner noted that the cause of death was by cynanide poisoning caused by eating an apple that was laced with cyanide.

Turing made outstanding contributions to mathematics, computer science, artificial intelligence and biology. The Association for Computing Machinery awards the annual "Turing Award" to individual who have made outstanding contributions to the computing field.

2.6 Shannon

Claude Shannon (Fig. 2.15) was born in Michigan in 1916. His primary degree was in Mathematics and Electrical Engineering at the University of Michigan in 1936. He was awarded a Ph.D. in mathematics from the Massachusetts Institute of Technology (MIT) in 1940. His initial work at MIT was to improve his supervisor's mechanical computing device known as the Differential Analyser by using electronic circuits rather than mechanical parts.

Shannon's research led to his insight that an electric circuit is similar to the Boolean Algebra that he had previously studied at the University of Michigan. The methods employed in the 1930s to design logic circuits were ad hoc. Shannon showed that Boolean algebra could be employed to simplify the design of circuits

Fig. 2.15 Claude Shannon

and telephone routing switches.[18] Shannon's results helped to renew interest in the work of George Boole on Boolean algebra, as it was now evident that the Boolean concept of true and false could be employed to represent the functions of switches in electronic circuits.

Shannon showed how to lay out circuitry according to Boolean principles, and his influential Masters thesis: "*A Symbolic Analysis of Relay and Switching Circuits*" [Sha:37] became the foundation for the practical design of digital circuits. These circuits are fundamental to the operation of modern computers and telecommunication systems, and Shannon's Masters Thesis is a key milestone in the development of modern computers. Shannon's insight of using the properties of electrical switches to do Boolean logic is the basic concept that underlies all electronic digital computers.

---

[18] Boolean algebra was able to simplify the arrangement of the electromechanical relays used in telephone switches.

He moved to the Mathematics Department at Bell Labs in the 1940s and commenced work that would lead to the foundation of modern Information Theory. Information includes messages that occur in any communications medium, for example, television, radio, telephone, and computers. The fundamental problem of communication is to reproduce at a destination point either exactly or approximately a message that has been sent from a source point. The problem is that information may be distorted by noise and the message received may differ from that which was originally sent.

Shannon provided a mathematical definition and framework for information theory that remains the standard today. The theory is described in "*A Mathematical Theory of Communication*" [Sha:48]. Shannon proposed the idea of converting any kind of data (e.g., pictures, sounds, or text) to binary digits: i.e., the data are reduced to binary bits of information. The information is then be transmitted over the communication medium. Errors or noise may be introduced during the transmission, and the objective is to reduce and correct errors. The received binary information is then converted back to the appropriate medium. Optimal communication of data is achieved by removing all randomness and redundancy.

Shannon is therefore the father of digital communication as used by computers, His theory was an immediate success with communications engineers and its impact was far reaching. Shannon's later work made a contribution to the field of cryptography, and this is described in "*Communication Theory of Secrecy Systems*" [Sha:49]. Shannon also invented a chess playing computer program in 1948. Shannon retired at the age of 50 and died aged 84 in 2001.

2.6.1 Boolean Algebra and Switching Circuits

Shannon showed that Boolean Algebra may be employed to design and analyse switching circuits. A circuit may be represented by a set of equations with the terms in the equations representing the various switches and relays in the circuit. He developed a calculus for manipulating the equations and this calculus is analogous to Boole's propositional logic. The design of a circuit consists of writing the algebraic equations that describe the circuit, and the equations are manipulated to yield the simplest circuit. The circuit may then be immediately drawn.

At any given time the switching circuit X_{ab} between two terminals a and b is either open (infinite impendence with $X_{ab} = 1$) or closed (zero impendence with $X_{ab} = 0$). The symbol 0 is employed to denote the hindrance of a closed circuit and the symbol 1 is used to denoted the hindrance of an open circuit. The expression X_{ab} is denoted pictorially in Fig. 2.16:

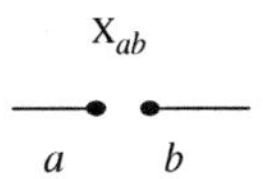

Fig. 2.16 Open circuit

For two circuits X_{ab} and Y_{cd} the expression $X_{ab} + Y_{cd}$ denotes the hindrance of the circuit formed when the circuit *ab* is joined serially to the circuit *cd*. Similarly, the expression $X_{ab} \bullet Y_{cd}$ denotes the hindrance of the circuit formed when the circuit *ab* is joined in parallel to the circuit *cd*. The circuit formed by joining two circuits serially is described pictorially in Fig. 2.17:

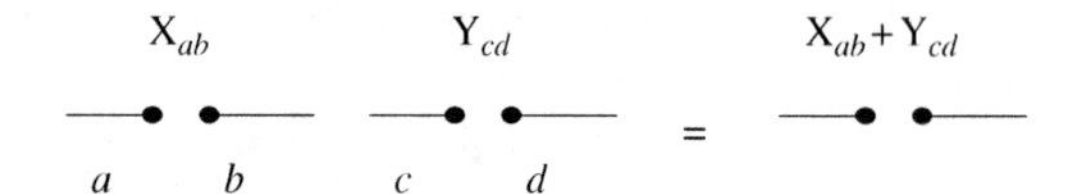

Fig. 2.17 Serial circuits

This will usually be written as X + Y as it is more common to write the symbols X,Y instead of X_{ab} and Y_{cd}. . Similarly, the circuit formed by joining two circuits in parallel is described pictorially in Fig. 2.18.

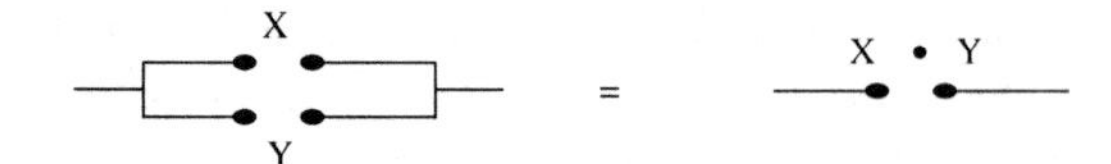

Fig. 2.18 Parallel circuits

2.6.1.1 Properties of Circuits

It is evident from the above definitions that the following laws are true (Table 2.8):

Table 2.8 Properties of circuits

Property name	Property	Interpretation
Idempotent Property	$0 \bullet 0 = 0$	A closed circuit in parallel with a closed circuit is a closed circuit.
	$1 + 1 = 1$	An open circuit in series with an open circuit is an open circuit.
Additive Identity (0)	$1 + 0 = 0 + 1 = 1$	An open circuit in series with a closed circuit (in either order) is an open circuit.
Multiplicative Identity (1)	$1 \bullet 0 = 0 \bullet 1 = 0$	A closed circuit in parallel with an open circuit (in either order) is a closed circuit.
Additive Identity (0)	$0 + 0 = 0$	A closed circuit in series with a closed circuit is a closed circuit.
Multiplicative Identity (1)	$1 \bullet 1 = 1$	An open circuit in parallel with an open circuit is an open circuit.

At any given time the circuit is either open or closed: i.e., $X = 0$ or $X = 1$. The following theorems may be proved in (Table 2.9):

Table 2.9 Properties of circuits (contd.)

Property name	Property
Commutative	$x + y = y + x$
Property	$x \bullet y = y \bullet x$
Associative	$x + (y + z) = (x + y) + z$
Property	$x \bullet (y \bullet z) = (x \bullet y) \bullet z$
Distributive	$x \bullet (y + z) = (x \bullet y) + (x \bullet y)$ $x + (y \bullet z) = (x + y) \bullet (x + z)$
Identity	$x + 0 = x = 0 + x$ $1 \bullet x = x \bullet 1 = x$
	$1 + x = x + 1 = 1$ $0 \bullet x = x \bullet 0 = 0$

The negation of a hindrance X is denoted by X′. It yields the following properties (Table 2.10).

Table 2.10 Properties of circuits (contd.)

Property name	Property
Negation	$X + X' = 1$
	$X \bullet X' = 0$
	$0' = 1$
	$1' = 0$
	$(X')' = X$
De Morgan's Law	$(X + Y)' = X' \bullet Y'$
	$(X \bullet Y)' = X' + Y'$
De Morgan's Law	$(X + Y + \ldots)' = X' \bullet Y' \bullet \ldots$
(Generalised)	$(X \bullet Y \bullet \ldots)' = X' + Y' + \ldots$

Functions of variables may also be defined in the calculus in terms of the "+", "•" and negation operations. For example, the function $f(X,Y,Z) = XY' + X'Z' + XZ'$ is an example of a function in the calculus and it describes a circuit.

2.6.1.2 Analogy Circuits and Boolean Logic

The calculus for circuits is analogous to Boole's propositional logic, and the analogy may be seen with the following Table 2.11:

Table 2.11 Circuits and Boolean algebra

Symbol	Digital circuits	Propositional logic
X	The circuit X	The proposition X
0	The circuit is closed	The proposition is false
1	The circuit is open	The proposition is true
X + Y	Series connection of circuits X and Y	The proposition which is true when either X or Y is true, and false otherwise
X • Y	Parallel connection of circuits X and Y	The proposition which is true when both X and Y are true.
X′	The circuit that is open when X is closed and vice versa	X′ is false when X is true and vice versa.
X = Y	The circuits open and close simultaneously	Each proposition implies the other

Any expression formed with the additive, multiplicative and negation operators forms a circuit containing serial and parallel connections only. To find the simplest circuit with the least number of contacts all that is required is to manipulate the mathematical expression into the form in which the fewest variables appear.

For example, the circuit represent by $f(X,Y,Z) = XY' + X'Z' + XZ'$ is given in Fig. 2.19:

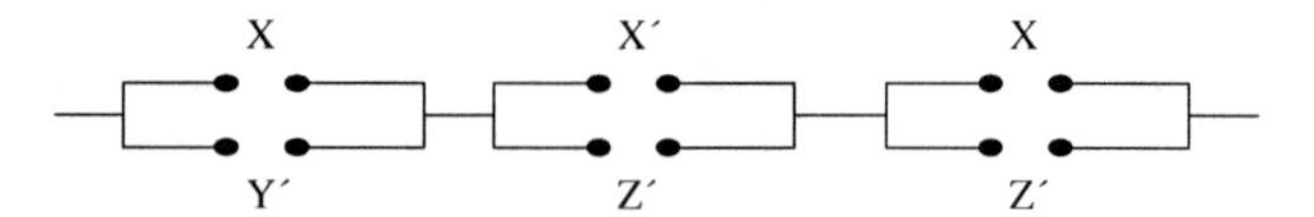

Fig. 2.19 Simplifying circuits

However, this circuit may be simplified by noting that (Fig. 2.20):

$$\begin{aligned} f(X,Y,Z) &= XY' + X'Z' + XZ' \\ &= XY' + (X' + X)Z' \\ &= XY' + 1Z' \\ &= XY' + Z' \end{aligned}$$

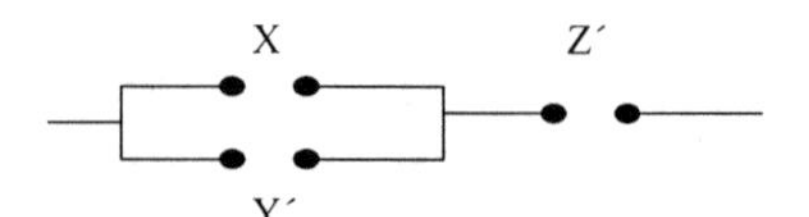

Fig. 2.20 Simplified circuit

2.6.2 Information Theory

Early pioneering work on Information Theory was done by Nyquist in the 1920s (Fig. 2.21). Shannon is recognised as the father of Information Theory due to his work in formulating a mathematical foundation for the field in his classic 1948 paper "A Mathematical Theory of Communication" [Sha:48]. The key problem to be solved in information theory is the reliable transmission of a message from a source point over a channel to a destination point.[19] The problem is that there may be noise in the channel that may distort the message being sent, and the engineer wishes to ensure that the message received has not been distorted by noise. The objective of information theory is to provide answers to how rapidly or reliably a

[19] The system designer may also place a device called an encoder between the source and the channel and a device called a decoder between the output of the channel and the destination.

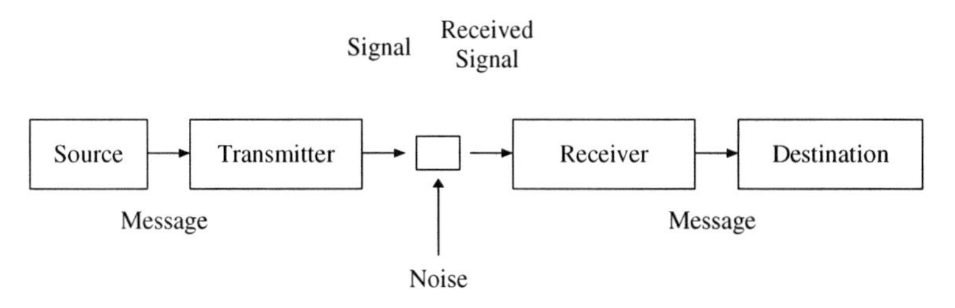

Fig. 2.21 Information theory

message may be sent from the source point to the destination point. The meanings of the messages are ignored as they are irrelevant from an engineering viewpoint.

Shannon's theory of information is based on probability theory and statistics. One important concept is that of entropy which measures the level of uncertainty associated with a random variable X. The amount of information in common between two random variables can be used to find the communication rate across a channel.

Shannon derived a fundamental limit theorem, and this theorem states the limitations on the reliability level that may be achieved for a given source and channel. Further, he showed that for large encoder delays, it is possible to achieve performance that is essentially as good as the fundamental limit. Shannon's noisy-channel coding theorem states that reliable communication is possible over noisy channels provided that the rate of communication is below a certain threshold called the "channel capacity". The channel capacity can be approached by using appropriate encoding and decoding systems. Shannon's work established the fundamental limits on communication.

Shannon's theory also showed how to design more efficient communication and storage systems.

2.6.3 Cryptography

Shannon is considered the father of modern cryptography with his influential 1949 paper "Communication Theory of Secrecy Systems" [Sha:49]. His work established a solid theoretical basis for cryptography and for cryptanalysis. Shannon defined the basic mathematical structures that underly secrecy systems, and a secrecy system is defined to be a transformation from the space of all messages to the space of all cryptograms. Each possible transformation corresponds to encryption with a particular key. The transformations are generally reversible as this allows the original message to be deciphered from the cryptogram provided that the key is known. The basic operation of a secrecy system is described in Fig. 2.22:

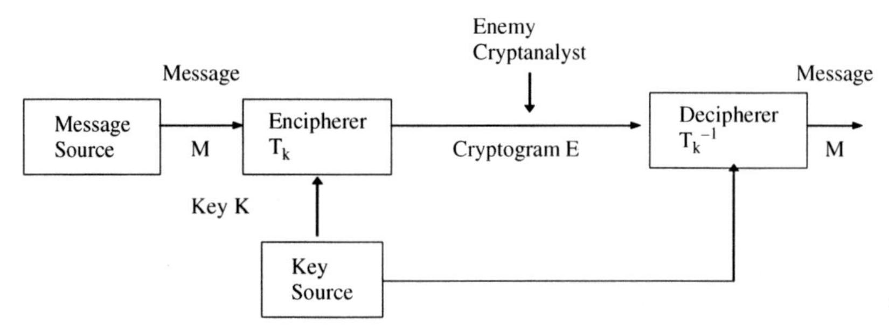

Fig. 2.22 Cryptography

First, the key is selected and sent to the receiving point. The choice of key determines the particular transformation of the message. The key is used to transform the chosen message into a cryptogram (i.e., the encrypted text). The cryptogram is then transmitted over a channel to the receiver who uses the key to apply the inverse transformation to decipher the cryptogram into the original message.

The enciphering of a message is a functional operation. Suppose M is a message, K the key and E is the encrypted message then:

$$E = f(M, K)$$

This is more often written as a function of one variable $E = T_i M$ where the index i corresponds to the particular key being used. It is assumed that there are a finite number of keys $K_1, \ldots K_m$ and a corresponding set of transformations $T_1, T_2, \ldots, T_m$. Each key has a probability p_i of being chosen as the key. The encryption of a message M with key K_i is therefore given by:

$$E = T_i M$$

Following the receipt of the encrypted message at the receiving end it is possible to retrieve the original message by:

$$M = T_i^{-1} E$$

The channel that the encrypted text is sent may be intercepted by an enemy who will try techniques to break the encrypted text. The enemy will examine the cryptogram and attempt to guess the key and the message from the cryptogram.

2.7 Von Neumann

John von Neumann (Fig. 2.23) was a Hungarian mathematician who made fundamental contributions to mathematics, set theory, computer science, economics, quantum analysis and many more. He was born into a non-practicing Jewish family in Budapest, Humgary in 1903. His father was a lawyer in a bank and the family

Fig. 2.23 John von Neumann, Los Alamos Courtesy United States Government Archives.

was well-connected. János (John) von Neumann was a child prodigy and displayed an extra-ordinary talent for mathematics at an early age. This included the ability to divide two 8-digit numbers at the age of six. By the age of twelve he was at the graduate level in mathematics. He attended the Lutheran Gymnasium which was one of the top schools in Hungary.

He received his PhD in Mathematics from the University of Budapest in 1926 and was awarded a diploma in chemical engineering from the Technical University of Zurich in the same year. His PhD degree was concerned with the axiomatization of set theory. Dedekind and Peano had axiomatized arithmetic in the late nineteenth century and and Hilbert had axiomatized geometry. However, set theory as developed by Cantor and formalized by Frege was dealt a major blow by Russell's paradox. Russell's paradox was discovered by Bertand Russell in 1901 and he communicated the paradox to Frege in 1902. Russell had posed the question whether the set S of all sets that do not contain themselves as members contains itself as a member: i.e., does $S \in S$? In either case, a contradiction arises since if $S \in S$ then as S is a set that does not contain itself as a member: i.e., $S \notin S$. Similarly, if $S \notin S$ then S is a set that does not contain itself as a member and therefore $S \in S$. Russell's paradox created a crisis in the foundations of set theory (and in the foundations of mathematics) in the early twentieth century.

Hilbert and the formalists dealt with Russell's paradox by insisting on the use of finite, well-defined and constructible objects. Further, they insisted on rules of inference that were absolutely certain. The axiomatization of set theory by Zermelo and Frankel in 1908 showed how the contradictions in set theory could be resolved by restricting set comprehension. The axiomatization employed prevents the construction of a set of all sets which do not belong to themselves. It allows the construction of all sets that are used in mathematics. However, their axiomatization did not exclude the possibility of the existence of sets which belong to themselves.

Russell resolved the paradox in his Principles of Mathematics by developing the "Theory of Types" which avoids self-reference. Russell argued that the set S (the set of all sets that are not members of themselves) can be avoided if all propositional sentences are arranged into a hierarchy. The lowest level of the hierarchy consists of sentences about individuals; the next level consists of sentences about sets of individuals; the next level is sets of sets of individuals, and so on. The level is essentially the type of the object. It is then possible to refer to all objects for which a given predicate holds only if they are all of the same "type". It is required to specify the objects that a predicate function will apply to before defining the predicate. For example, before the predicate *prime number* is defined it is first required to specify the objects that the predicate will apply to: i.e., the set of natural numbers. The objects that a predicate may be applied will not include any object that is defined in terms of the predicate itself (as they are of different types).

Von Neumann's contribution in his PhD thesis was to show how the contradiction in set theory can be avoided in two ways. One approach is to employ the axiom of foundation, and the other approach is to employ the concept of a class. The axiom of foundation specifies that every set can be constructed from the bottom up in a sequence of steps such that if one set belongs to another then it must come before the second in the sequence. He showed that the addition of this new axiom to the other axioms of set theory did not produce a contradiction. His second approach was to employ the concept of a class, and to define a set as a class which belongs to other classes. A proper class is a class that does not belong to any other class. The class of all sets which do not belong to themselves can be constructed. However, it is a proper class and is not a set.

Von Neumann next considered the problem of the axiomatization of quantum theory. The axiomatization of physical theories was listed as one of the key problems to be solved by Hilbert in 1900. Hilbert had presented his famous list of twenty-three problems at the international congress of mathematics in 1900. Quantum mechanics had deep foundational and philosophical issues especially with respect to non-determinism. The two existing formulations for quantum mechanics were the wave mechanical formulation by Schrödinger, and the matrix mechanical formulation by Heisenberg. However, there was no single theoretical formulation.

Von Neumann developed an elegant mathematical theory that showed how the physics of Quantum Theory could be reduced to the mathematics of linear operators on Hilbert spaces [VN:32]. The theory formulated by Von Neumann included the Schödinger and Heisenberg formulations as special cases. Heisenberg's Uncertainty principle with respect to position and the momentum of microscopic particles was translated to the non-commutativity of two operators. Von Neumann's work in Quantum Mechanics was very influential. However, many physicists preferred an alternate approach formulated by Dirac which involved the use of the Diral Delta function. Von Neumann and Birkhoff later proved that quantum mechanics requires a logic that is quite different from classical logic. For example, the commutativity of Boolean algebra is no longer valid. That is, $A \wedge B \neq B \wedge A$. Similarly, the distributive laws are no longer valid.

Von Neumann was a private lecturer in Germany between 1926 and 1930. He moved to the United States in the early 1930s to take a position as professor of mathematics at the newly formed Institute for Advanced Studies at Princeton in New Jersey. He became an American citizen in 1937 and remained at Princeton for the rest of his life.

Von Neumann made contributions to Economics and showed how game theory and the theory of equilibria could be used in Economics. His minimax theorem shows that in certain zero sum games there exists a strategy to minimize their maximum losses. He later extended the theorem to include games involving more than two players. Another important contribution made by von Neumann was to the problem of the existence of equilibrium in mathematical models of supply and demand. His solution employed the Browser's fixpoint theorem.

Von Neumann became very interested in practical problems and was involved in several consultancies including the United States navy, IBM, the CIA and the Rand Corporation. He became an expert in the field of explosions and discovered that large bombs are more devastating if they explode before touching the ground. This result was employed with devastating effect in the bombing of Hiroshima and Nagasaki in 1945 at the altitude calculated by Von Neumann to inflict maximum damage. He contributed to the development of the Hydrogen bomb and to improved methods to utilize nuclear energy. Von Neumann and Ulam also developed computer simulations to assist in the development of the bomb.

He gave his name to the von Neumann architecture used in almost all computers. However, credit should also be given to Eckhert and Mauchly who were working on this concept during their work on ENIAC. Virtually every computer is a von Neumann machine. He also created the field of cellular automata. He is also credited as the inventor of the merge-sort algorithm (in which the first and second halves of an array are each sorted recursively and then merged). He also invented the Monte Carlo method that allows complicated problems to be approximated through the use of random numbers.

He was strongly anti-communist and worked on various military committees developing various scenarios of nuclear proliferation including intercontinental and submarine missiles.

He died at the relatively young age of 54 in 1957, and his death was the result of cancer possibly obtained due to exposure to radiation from atomic bomb tests in the pacific. There is an annual IEEE von Neumann medal awarded by IEEE for outstanding achievements in computer science.

2.7.1 Von Neumann Architecture

The earliest computing machines had fixed programs that were designed to do a specific task. For example, Babbage's difference engine and the tabulating machines developed by Hermann Hollerith were fixed program computers. The limitation of a fixed program computer is that it is designed and programmed to do a specific task (or tasks). If it is required to change the program of such a machine then it is usually necessary to re-wire and re-design the machine. That is, the early machines were designed for the particular problem that needed to be solved. In contrast, today's computers are more general purpose, and are designed to allow a variety of programs to be run on the machine. The process of re-designing the early computers was a complex manual process, and involved engineering designs and physical changes.

Von Neumann architecture is a computer design that uses a single store for both machine instructions and programs. It is also known as a stored program computer and it is a sequential architecture.

Virtually every computer is a von Neumann machine and von Neumann is credited with inventing this architecture. However, Eckhert and Mauchly also contributed to the architecture as they were working on this concept during their work on ENIAC, and in designing its successor EDVAC. The ENIAC was a machine that had to be physically rewired in order to perform different tasks, and it was clear to the team that there was a need for an architecture that would allow a machine to perform different tasks without physical rewiring each time.

Von Neumann architecture arose on work done by von Neumann, Eckhert, Mauchy and others on the design of EDVAC which was the successor to ENIAC. Von Neumann was working as a consultant to the United States. Army Ordnance Department at the time, and his draft report on EDVAC [VN:45] described the new architecture. EDVAC itself was built in 1949. The key components of von Neumann architecture include (Table 2.12):

Table 2.12 Von Neumann architecture

Component	Description
Arithmetic Unit	The arithmetic unit is capable of performing basic arithmetic operations.
Control Unit	Executes the instructions that are stored in memory.
	The program counter contains the address of the next instruction to be executed, and this instruction is fetched from memory and executed. That is, the basic cycle is Fetch and Execute (as in Fig. 2.24) and the control unit contains a built in set of machine instructions.
Input–Output Unit	The Input and output unit allows the computer to interact with the outside world.
	The input of what needs to be computed is entered, and the computed results are printed or displayed.
Memory	There is a one dimensional memory that stores all of the program instructions and data.
	These are usually kept in different areas of memory.
	The memory may be written to or read from: i.e., it is random access memory (RAM).
	The program itself and the data reside in the same memory.
	The program instructions are binary values, and the control unit decodes the binary value to determine the particular instruction to execute.

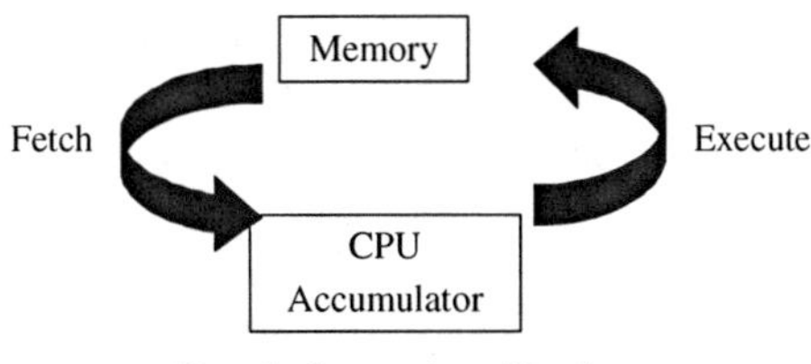

Fig. 2.24 Fetch-execute cycle

The key approach to building a general purpose device according to von Neumann was in its ability to store not only its data and the intermediate results of computation, but also to store the instructions, or commands for the computation. The computer instructions can be part of the hardware for specialized machines, but for general purpose machines the computer instructions must be as changeable as the data that is acted upon by the instructions. Von Neumann's insight was to recognize that both the machine instructions and data could be stored in the same memory.

The key advantage of the new von Neumann architecture was that it was much simpler to reconfigure a computer to compute a different task. All that was required was to put the new machine instructions in computer memory rather than physically rewiring a machine that was required with ENIAC (Fig. 2.25).

The generally accepted first computer with a von Neumann architecture was the Manchester Mark I computer.[20] It was designed and built at Manchester University in England by Frederic Williams, Tom Kilburn and others. Their first prototype Mark 1 computer was called "Baby" (Fig. 2.26), and it ran its first stored program in 1948.

Fig. 2.25 Two women working with ENIAC computer
United States Army Photo.

[20] Strictly speaking, Zuse's Z3 which was built in 1941 actually predated the Manchester Mark 1. Even though the Z3 was not an electronic computer it was programmable by punched film.

Fig. 2.26 Replica of the Manchester Baby
Courtesy of Tommy Thomas

This machine demonstrated the feasibility and potential of the stored program. The memory of the machine consisted of 32 32-bit words, and it took 1.2 milliseconds to execute one instruction: i.e., 0.00083 MIPS. This is very slow compared to today's computers which are rated at speeds of up to 1000 MIPs. The team in Manchester developed the machine further, and in 1949, the Manchester Mark 1 (Fig. 2.27) was available.

Fig. 2.27 The Manchester Mark 1 computer
Courtesy of the School of Computer Science. The University of Manchester.

Ferranti (a British company) produced the world's first commercial computer by building on the Manchester Mark 1. Their first machine was called the Ferranti Mark 1, and it was released in 1951.

There is more emphasis today on architectures that are free from the constraints of the traditional von Neumann architecture. The goals are to have an architecture that will increase productivity, and in recent years a number of computers have claimed to be "non-von Neumann". This trend is likely to continue as new architectures evolve, and where both hardware and software are freed from the limitations of the von Neumann architecture. The limitations include:

- Limited to sequential processing
- Not very well suited to parallel processing (multiple data/instructions)

2.8 Konrad Zuse

Konrad Zuse (Fig. 2.28) was a German engineer who was born in Bonn in 1910. He worked initially as a design engineer at the Henschel aircraft factory in eastern Germany, but resigned from his job after about a year to commence work on building a program driven and programmable machine at his parents' apartment. He produced his first machine called the Z1 in 1936, and this machine was a binary electrically driven mechanical calculator with limited programmability. The machine was capable of executing instructions read from the program punch cards only, but the program itself was never loaded into the memory.

The machine was essentially a 22-bit floating point value adder and subtracter. The input to the machine was done with a decimal keyboard and the output was decimal digits. The machine included some control logic which allowed it to perform more complex operations such as multiplications and division. These operations were performed by repeated additions for multiplication, and repeated subtractions for division. Multiplication took approximately 5 seconds. The computer memory contained 64 22-bit words. Each word of memory could be read from and written to by the program punch cards and the control unit. It had a clock speed of 1 Hz, and two floating point registers of 22 bits each. The machine itself was unreliable and a reconstruction of it is in the Deutch Technikmuseum in Berlin.

His next attempt was the creation of the Z2 machine. This was a mechanical and relay computer created in 1939 and it aimed to improve on the Z1. It used a similar mechanical memory but replaced the arithmetic and control logic with electrical relay circuits. It used 16-bit fixed point arithmetics instead of the 22-bit used in the Z1. It had 32 bytes of memory (16 words of 16 bits) and had a clock speed of 3 Hz.

His greatest achievement was the completion of the first functional tape-stored-program-controlled computer, the Z3, in 1941. The Z3 was proven to be Turing-complete in 1998. It was only discovered in the aftermath of world-war two that

Fig. 2.28 Konrad Zuse
Courtesy of Horst Zuse, Berlin.

a program controlled computer called the Z3 had been completed in Germany in 1941. It pre-dates the Harvard Mark I[21] and the Manchester Mark 1.

The Z3 was very sophisticated for its time and it was based on relays. It used the binary number system and could perform floating-point arithmetic. It had a clock speed of 5Hz and multiplication and division took 3 seconds. The input to the machine was with a decimal keyboard and the output was on lamps that could display decimal numbers. The word length was 22-bits. There is a reconstruction of the Z3 computer in the Deutch Museum in Munich. Zuse also developed the Plankalkül high-level programming language in 1946.

[21] The Harvard Mark I was a large-scale automatic digital computer developed by Howard H. Aiken (and team) and built by IBM in America between 1939 and 1944. IBM marketed it as the ASCC machine. Grace Murray Hopper was a programmer on the Mark 1.

Zuse was completely unaware of any computer-related developments in Germany or in other countries until a very late stage. He independently conceived and implemented the principles of modern digital computers in isolation.

2.9 Review Questions

1. Describe Boolean Algebra.
2. Describe the Difference Engine.
3. Describe the Analytic Engine.
4. Discuss the Turing Machine.
5. Discuss the Turing Test.
6. Discuss the Enigma Machine and the work done at Bletchey Park.
7. Discuss the application of Boolean Algebra to Switching Circuits.
8. Discuss Information Theory.
9. Discuss Shannon's contribution to Cryptology.
10. Discuss the Von Neumann Architecture.

2.10 Summary

This chapter considered the contributions of important yures in the history of computing including Boole, Babbage, Shannon, Turing and von Neumann. George Boole was an English mathematician who developed Boolean Algebra which is the foundation of all modern computers.

Charles Babbage did pioneering work on the Difference Engine. This was essentially a mechanical calculator (analogous to modern electronic calculators), and it was designed to compute polynomial functions. He also designed the Analytic Engine, and this was the design of the world's first mechanical computer. It included a processor, memory, and a way to input information and output results.

Turing is famous for his work on a theoretical mathematical machine termed the "Turing Machine", and he proved that anything that is computable is computable by this theoretical machine. He also made contributions to the cryptography and to Artificial Intelligence. He devised the famous "Turing Test" that is a test of machine intelligence.

Claude Shannon was an American mathematician and engineer who made fundamental contributions to computing. He was the first person to see the applicability of Boolean Algebra to simplify the design of circuits and telephone routing switches, and his insight of using the properties of electrical switches to do Boolean logic is the basic concept that underlies all electronic digital computers. His later work at Bell Labs laid the foundation of modern Information Theory and modern Cryptography.

Von Neumann was a Hungarian mathematician who made fundamental contributions to mathematics, physics and computer science. He gave his name to the von Neumann architecture that is used in almost all computers.

Zuse was a German engineer who developed the Z3 machine in 1941. He also developed the Plankalkül high-level programming language in 1946.

Chapter 3
Computer Programming Languages

Key Topics

Generations of Programming Languages
Imperative Languages
ALGOL
Fortran and Cobol
Pascal and C
Object-Oriented Languages
Functional Programming Languages
Logic Programming Languages
Syntax and Semantics

3.1 Introduction

Computer programming languages have evolved from the early days of computing. The earliest programming languages used machine code to instruct the computer, and the next development was to use low-level assembly languages to represent machine language instructions. These were then translated to the machine code by an assembler. The next step was to develop high-level programming languages such as Fortran and Cobol.

A first-generation programming language (or 1GL) is a machine-level programming language that consists of 1s and 0s. Their main advantage was execution speed and efficiency. They could be directly executed on the central processing unit (CPU) if the computer, and there is no need to employ a compiler or assembler to translate from a high-level language or assembly language to the machine code.

However, writing in a machine language was difficult and error prone as it involved writing a stream of binary numbers. This made the programming language difficult to learn and difficult to correct should any errors occur. The programming instructions were entered through the front panel switches of the computer system, and adding new code was difficult. Further, the machine code was not portable as the machine language for one computer could differ significantly from that of another computer. Often, the program needed to be totally re-written for the new computer.

First generation languages are mainly of historical interest today, and were mainly used on the early computers. A program written in a high-level programming language is generally translated by the compiler[1] into the machine language of the target computer for execution.

Second generation languages, or 2GL, are low-level assembly languages that are specific to a particular computer and processor. However, assembly languages are unlike first-generation programming languages in that the assembly code can be read and written more easily by a human. They require considerably more programming effort than high-level programming languages, and they are more difficult to use for larger applications. The assembly code must be converted into the actual machine code in order to run on the computer. The conversion of the assembly code to the machine code is simply a mapping of the assembly language code into the binary machine code (i.e., the first generation language). The assembly language is specific to a particular processor family and environment, and it is therefore not portable.

A program written in assembly language for a particular processor family needs to be re-written for a different platform. However, since the assembly language is in the native language of the processor it has significant speed advantages over high-level languages. Second generation languages are still used today, but they have generally been replaced by high-level programming languages.

The third generation languages, or 3GL, are high-level programming languages such as Pascal, C or Fortran. They are general purpose languages and have been applied to business, scientific and general applications. They are designed to be easier for a human to understand and include features such as named variables, conditional statements, iterative statements, assignment statements, and data structures. Early examples of third generation languages are Fortran, ALGOL and COBOL. Later examples are languages such as C, C++ and Java. The advantages of these high-level languages over the older second generation languages were:

- Ease of readability
- Clearly defined syntax (and semantics[2])

[1] This is true of code generated by native compilers. Other compilers may compile the source code to the object code of a Virtual Machine, and the translator module of the Virtual Machine translates each byte code of the Virtual Machine to the corresponding native machine instruction. That is, the Virtual Machine translates each generalised machine instruction into a specific machine instruction (or instructions) that may then be executed by the processor on the target computer. Most computer languages such as C require a separate compiler for each computer platform (i.e., computer and operating system). However, a language such as Java comes with a virtual machine for each platform. This allows the source code statements in these programs to be compiled just once, and they will then run on any platform.

[2] The study of programming language semantics commenced in the 1960s. It includes work done by Hoare on Axiomatic Semantics; work done by Bjørner and Jones at IBM in Vienna on Operational Semantics; and work done by Scott and Strachey on Denotational Semantics.

- Suitable for Business or Scientific applications
- Machine independent
- Portability to other platforms
- Ease of debugging
- Execution speed.

These languages are machine independent and may be compiled for different platforms. Professional programmers employ these languages to develop almost any type of program. The early 3GLs were procedure-oriented languages that required programmers to solve programming problems by writing the program instructions in the sequence in which they must be executed to solve the problem. The later 3GLs were object-oriented[3] and the programming tasks were divided into objects. These objects may be employed to build larger programs, in a manner that is analogous to building a prefabricated building. Examples of modern object-oriented language are the Java language that is used to build web applications, C++ and Smalltalk.

High-level programming languages allow programmers to focus on problem solving rather than on the low level details associated with assembly languages. They are easier to debug and easier to maintain than assembly languages. Most of the third generation languages are procedural in that they focus on how something is done rather than on what needs to be done. They include von Neumann constructs such as sequential statements, conditional (or branch) statements, and iterative statements.

The fourth generation languages, or 4GLs, are languages that consist of statements similar to a human language. Most fourth generation languages are not procedural, and are often used in database programming. The term was coined in the early 1980s in a book by James Martin[4] [Mar:82]. They specify what needs to be done rather than how it should be done. All 4GLs are designed to reduce programming effort (e.g., reducing the time that to develop the software), as this helps to reduce the cost of software development. However, they are not always successful in this task, and sometimes the resultant code is difficult to maintain. Further, 4GLs are slow when compared to compiled languages.

There are various kinds of 4GLs including report generators and form generators. Report generators take a description of the data format and the report that is to be created, and then automatically generate a program to produce the report. Form generators are used to generate programs to manage online interactions with the application system users. There are more sophisticated 4GLs that attempt to generate whole systems from specifications of screens and reports.

A fifth-generation programming language, or 5GL, is a programming language that is based around solving problems using constraints applied to the program,

---

[3] Object-oriented programming was originally developed by Norwegian Research with Simula-67 in the late 1960s.

[4] James Martin is a consultant and author and has written many books on information technology. He was an early promoter of fourth generation languages and has written extensively on CASE and on Rapid Application Development. There is a James Martin twenty-first century school at Oxford.

rather than using an algorithm written by the programmer. Fifth-generation languages are designed to make the computer (rather than the programmer) solve the problem for you. The programmer only needs to be concerned with the specification of the problem and the constraints to be satisfied, and does not need to be concerned with the algorithm or implementation details. These languages are mainly used for research purposes especially in the field of artificial intelligence. Prolog is one of the best known fifth generation languages, and it is a logic programming language.

Early research on fifth generation languages was encouraging, and it was predicted that they would eventually replace all other languages. However, the task of deriving an efficient algorithm from a set of constraints for a particular problem is non-trivial. This step has not been successfully automated and still requires the insight of a human programmer. Fifth-generation languages are used mainly in academia.

3.2 Early Programming Languages

One of the earliest programming languages developed was Plankalkül. This high-level imperative programming language was developed by Konrad Zuse in 1946. Zuse was a German engineer who designed the Z1, Z2 and Z3 computers. The name "Plankalkül" derives from "Plan" and "Kalkül" (meaning calculus) and essentially means a calculus for a computing plan, or a calculus of programs.

Plankalkül is a relatively modern language, and this is surprising for a language developed in 1946. There were no compilers for programming languages in 1946, and physical rewiring of the machine were required. For example, the ENIAC needed to be rewired for every different problem. Plankalkül employed data structures and Boolean algebra and included a mechanism to define more powerful data structures. Zuse demonstrated that the Plankalkül language could be used to solve scientific and engineering problems, and he wrote several example programs including programs for sorting lists and searching a list for a particular entry.

Zuse completed the definition of Plankalkül in 1946, but as there was no compiler for the language it was not possible to execute any Plankalkül programs. The language was finally published in 1972, and a compiler for the language was eventually developed in 2000 at the Free University of Berlin. The compiler was developed in Java for a subset of the Plankalkül language, and the first programs were run over 55 years after the conception of the language. The main features of Plankalkül are:

- It is a high-level language
- Its fundamental data types are arrays and tuples of arrays
- There is a While construct for iteration
- Conditionals are addressed using guarded commands
- There is no GOTO statement
- Programs are non-recursive functions
- Type of a variable is specified when it is used

The main constructs of the language are variable assignment, arithmetical and logical operations, guarded commands and while loops. There are also some list and set processing functions.

3.3 Imperative Programming Languages

Imperative programming is a programming style that describes computation in terms of a program state, and statements that change the program state. The term "imperative" in a natural language such as English, is a command to carry out a specific instruction or action. Similarly, imperative programming consists of a set of commands to be executed on the computer, and is therefore concerned with *how* the program will be executed. The execution of an imperative command generally results in a change of state.

Imperative programming languages are quite distinct from functional and logical programming languages. Functional programming languages, like Miranda, have no global state, and programs consist of mathematical functions that have no side affects. In other words, there is no change of state, and the variable x will have the same value later in the program as it does earlier. Logical programming languages, like Prolog, define *what* is to be computed, rather than *how* the computation is to take place. Most commercial programming languages are imperative languages, with interest in functional programming languages and relational programming languages being mainly academic. Imperative programs tend to be more difficult to reason about due to the change of state.

High-level imperative languages use program variables, and employ commands such as assignment statements, conditional commands, iterative commands, and calls to procedures. An assignment statement performs an operation on information located in memory, and stores the results in memory. The effect of an assignment statement is usually to change the program state. A conditional statement allows a statement to be executed only if a specified condition is satisfied. Iterative statements allow a statement (or group of statements) to be executed a number of times. The compiler converts the source code of the high-level programming language into the machine code of the computer. The machine code is then executed on the hardware of the computer. A change of state is, in effect, a change to the contents of memory, and the machine code of the computer has instructions to modify the contents of memory as well as commands to jump to a location in memory. Assembly languages and machine code are imperative languages.

High-level imperative languages allow the evaluation of complex expressions. These may consist of arithmetic operations and function evaluations, and the resulting value of the expression is assigned to memory. The earliest imperative languages were machine code.

FORTRAN was developed in the mid-1950s, and it was one of the earliest programming languages. ALGOL was developed in the late 1950s and 1960s, and it became a popular language for the expression of algorithms. *COBOL* was designed in the late 1950s as a programming language for business use. BASIC (Beginner's All Purpose Symbolic Instruction Code) was designed by George Kemeny and Tom Kurtas in 1963 as a teaching tool. Pascal was developed in the early 1970s as a teaching language by the Swiss computer scientist, Niklaus Wirth. Wirth received a Turing award in 1984 for his contributions to programming languages and software engineering. The C programming language was developed in the early 1970s at Bell Laboratories in the United States. It was designed as a systems programming language for the Unix operating system.

Jean Ichbiah and others at Honeywell began designing Ada for the United States military in 1974, and the specification of the language was completed in 1983. Object-oriented programming became popular in the 1980s. These languages were imperative in style, but included features to support objects. Bjarne Stroustrup designed an object-oriented extension of the C language called C++, and this was first implemented in 1985. Java was released by Sun Microsystems in 1996.

3.3.1 Fortran and Cobol

FORTRAN (FORmula TRANslator) was developed by John Backus at IBM in the mid-1950s for the IBM 704 computer. It is a compiled high-level imperative language and includes named variables, complex expressions, and subprograms. It was designed for scientific and engineering applications, and it became the most important programming language for these applications. There have been many versions of Fortran since then, including Fortran II, Fortran IV, Fortran 66, Fortran 77 and Fortran 90. The first Fortran compiler was available in 1957. The main statements of the language include:

- Assignment Statements (using the = symbol)
- IF Statements
- Goto Statements
- DO Loops

Fortran II was developed in 1958, and it introduced sub-programs and functions to support procedural (or imperative) programming. Imperative programming specifies the steps the program must perform, including the concept of procedure calls, where each procedure (or subroutine) contains computational steps to be carried out. Any given procedure might be called at any point during a program's execution, including calls by other procedures or by itself. However, recursion was not allowed until Fortran 90. Fortran 2003 provides support for object-oriented programming.

The basic types supported in Fortran include Boolean, Integer, and Real. Support for double precision and complex numbers was added later. The language included relational operators for equality (.EQ.), less than (.LT.), and so on. It was good at handling numbers and computation, and this was especially useful for mathematical and engineering problems. The following code (written in Fortran 77) gives a flavour of the language.

```
   PROGRAM HELLOWORLD

C FORTRAN 77 SOURCE CODE COMMENTS FOR HELLOWORLD
      PRINT '(A)', 'HELLO WORLD'
      STOP
    END
```

Fortran remains a popular programming language for some of the most intensive supercomputing tasks. These include climate modeling, simulations of the solar system, modelling the trajectories of artificial satellites, and simulation of automobile crash dynamics. It was initially weak at handling input and output, which was important to business computing.

This led to the development of the COBOL programming language by Grace Murray Hopper[5] and others in the late 1950s (Fig. 3.1). The Common Business Oriented Language (COBOL) was the first business programming language, and it was introduced in 1959. It was developed by a group of computer professionals called the Conference on Data Systems Languages (CODASYL).

The objective of the language was to improve the readability of software source code, and the statements in the language are similar to English. It has a very English-like grammar, and it was designed to make it easier for the average business user to learn the language. The only data types in the language were numbers and strings of text. The language allowed for these to be grouped into arrays and records, so that

Fig. 3.1 Grace Murray Hopper United States Government archives.

[5] Mary Hopper was a programmer on the Mark 1, Mark II and Mark III and UNIVAC 1 computers. She was the technical advisor to the CODASYL committee.

data could be tracked and organized better. For example, the operation of division is performed by the verbose statement:

```
'DIVIDE A BY B GIVING C REMAINDER D'.
```

In PASCAL, the equivalent statement would be:

```
C := A div B

D := A mod B
```

COBOL was the first computer language whose use was mandated by the US Department of Defense. The language remains in use today (almost 50 years later), and there is also an object-oriented version of COBOL.

3.3.2 ALGOL

ALGOL (short for ALGOrithmic Language) is a family of imperative programming languages. It was originally developed in the mid-1950s and was subsequently revised in ALGOL 60, and ALGOL 68. The language was designed by a committee of American and European computer scientists to address some of the problems with the Fortran programming language. It had a significant influence on later programming language design. ALGOL 60 [Nau:60] was the most popular member of the family, and Edsger Dijkstra developed an early ALGOL 60 compiler. John Backus developed a method for describing the syntax of the ALGOL 58 programming language. This was revised and expanded by Peter Naur for ALGOL 60, and the resulting approach is known as Backus Naur form (or BNF). The introduction of BNF for defining the syntax or programming languages was a major contribution of ALGOL-60.

ALGOL includes data structures and block structures. Block structures were designed to allow blocks of statements to be created (e.g., for procedures or functions). A variable defined within a block may be used within the block but is out of scope outside of the block.

ALGOL 60 introduced two ways of passing parameters to subprograms, and these are *call by value* and *call by name*. The call by value parameter passing technique involves the evaluation of the arguments of a function or procedure before the function or procedure is entered. The values of the arguments are passed to the function or procedure, and any changes to the arguments within the called function or procedure have no effect on the actual arguments. The call by name parameter passing technique is the default parameter passing technique in ALGOL 60. This involves re-evaluating the actual parameter expression each time the formal parameter is read. Call by name is used today in C/C++ macro expansion.

ALGOL 60 includes conditional statements (if ... then ... else) and iterative statements. It also includes the concept of recursion that allows a function or procedure to call itself. ALGOL has a relatively small number of basic constructs and a set of rules for combining those constructs. Other features of the language include:

- Dynamic arrays: These are arrays in which the subscript range is specified by variables. The size of the array is set at the time that storage is allocated.
- Reserved words: These are keywords that are not allowed to be used as identifiers by the programmer.
- User defined data types: These allow the user to design their own data types that fit particular problems closely.
- ALGOL uses bracketed statement blocks and it was the first language to use *begin end* pairs for delimiting blocks.

ALGOL was used mainly by researchers in the United States and Europe. There was a lack of interest to its adoption by commercial companies due to the absence of standard input and output facilities in its description. ALGOL 60 became the standard for the publication of algorithms, and it had a major influence on later language development.

ALGOL evolved during the 1960s but not in the right direction. The ALGOL 68 committee decided on a very complex design rather than the simple and elegent ALGOL 60 specification. The eminent computer scientist, C.A.R. Hoare, remarked that:

Comment 3.1 (Hoare on ALGOL 68) ALGOL 60 was a great improvement on its successors.

ALGOL 60 inspired many languages that followed it such as Pascal, C, Modula 2 and Ada.

3.3.3 Pascal and C

The Pascal programming language was developed by Niklaus Wirth in the early 1970s (Fig. 3.2). It is named after Blaise Pascal who was a well-known seventeenth century French mathematician and philosopher. Pascal was based upon the ALGOL programming language, and it was intended mainly to be used as a teaching language and to teach students structured programming.

Structured programming [Dij:68] was concerned with rigorous techniques to design and develop programs. Dijkstra argued against the use of the GOTO statement "GOTO Statement considered harmful" [Dij:68]. Today, it is agreed that the GOTO statement should only be used in rare circumstances.

The debate on structured programming [Dij:72] was quite intense in the late 1960s. This influenced language design and led to several languages that did not include the GOTO statement. The Pascal language includes the following constructs:

Fig. 3.2 Niklaus Wirth
Photo courtesy of Wikipedia.

- If Statement

```
if (A = B) then
        A := A + 1
else
        A := A - 1;
```

- While Statement

```
while A > B do
        A := A - 1;
```

- For Statement

```
for I := 1 to 5 do
        WRITELN(I);
```

- Repeat Statement[6]

```
repeat
        A := A + 1
until A > 5;
```

- Assignment Statement

```
I := I + 1;
```

- Case Statement[7]

```
case expr of
expr-value-1:
        R := R + 1;
expr-value-2:
        R := 4
end
```

---

[6] The Repeat Statement and the While Statement are similar. The key difference is that the statement in the body of the Repeat statement is executed at least once, whereas the statement within the body of a while statement may never be executed.

[7] The Case statement is, in effect, a generalised if statement.

The Pascal language has several reserved words (known as keywords) that have a special meaning within the language. These keywords may not be used as program identifiers. Examples of reserved words in Pascal include "begin, end, while, for, if", and so on. The following is an example of a simple Pascal program that displays 'Hello World':

```
program HELLOWORLD (OUTPUT);

begin
WRITELN ('Hello World!')
end.
```

Pascal includes several simple data types such as Boolean, Integer, Character and Real data types, and it also allows more advanced data types. The advanced data types include arrays, enumeration types, ordinal types, and pointer data types. The language allows the construction of more complex data types (known as records) from existing data types. Types are introduced by the reserved word "type".

```
type
c = record
      a: integer;
      b: char
      end;
```

Pascal includes a "pointer" data type, and this data type allows linked lists to be created by including a pointer type field in the record.

```
type
BPTR = B̂;
B = record
A : integer;
C : BPTR
end;
var
LINKLIST : BPTR;
```

Here the variable LINKLIST is a pointer to the data type B, where B is a record. The following statements create a new record and assigns the value 5 to the field A in the record.

```
NEW (LINKLIST);
LINKLIST^.A := 5;
LINKLIST^.C := nil;
```

Pascal is a block structured language, and Pascal programs are structured into procedures and functions. Procedures and functions can nest to any depth, and recursion (i.e., where a function or procedure may call itself) is allowed. Each procedure or function block can have its own constants, types, variables, and other procedures and functions. These are all within scope within the block.

There were a number of criticisms made of the Pascal programming language by Brian Kernighan in [Ker:81]. He argued that Pascal was a teaching language and that it was not suitable for serious programming. The deficiencies listed by Kernighan were addressed in later versions of the language. However, by then a new language called C had been developed at Bell Laboratories in the United States, and this language was to become very popular in industry. The C language is a general purpose and a systems programming language. It was developed by Dennis Ritchie and Ken Thompson in the early 1970s, and it was based on two earlier languages namely B and BCPL.

The C language was originally designed to write the kernel for the UNIX operating system, and C and UNIX often go hand in hand. Assembly languages had traditionally been used for systems programming, and the success of C in writing the kernel of UNIX led to its use on several other operating systems such as Windows and Linux. C also influenced later language development (especially the development of C++). It is the most commonly used programming language for system programming, and it is also used for application development. It includes a number of the features of Pascal. The programmer's bible for the original version of C is in [KeR:78], and the later ANSI definition of the language is in [KeR:88].

The original design goals of C were to provide low level access to memory, and efficient compilation with only a few machine language instructions for each of its core language elements. The language provided low-level capabilities, and it also aimed to assist with portability and machine-independence. A program that is written with portability in mind, and is compliant with respect to the ANSI C standards, may be compiled for a very wide variety of computer platforms and operating systems with minimal changes to its source code. The C language is now available on a wide range of platforms.

C is a procedural programming language and includes conditional statements such as the if statement, the switch statement, iterative statements such as the while statement or do statement, and the assignment statement.

- If Statement

```
if (A == B)
A = A + 1;
else
A = A - 1;
```
[8]

[8] The semi-colon in Pascal is used as a statement separator, whereas it is used as a statement terminator in C.

- While Statement

```
while (A > B)
A = A - 1;
```

- Do Statement[9]

```
do
A := A + 1;
while (A < 5);
```

- Assignment Statement

```
i = i + 1;
```

- Switch Statement[10]

```
switch (expr)
{
case label1:
    A = 3;
    break;[11]
case label2:
    B = 2;
    break;
default:
    A = B + 1;
    break;
}
```

One of the first programs that people write in C is the Hello world program.[12] This program displays Hello World on the screen.

[9] The Do Statement is similar to Pascal's Repeat statement, and the statement in the body of the Do statement is executed at least once.

[10] The Switch statement is similar to Pascal's Case statement. The main difference is that C includes a default case which is executed if none of the other labels are satisfied.

[11] The break statement causes an exit from the nearest enclosing switch statement.

[12] The Hello world program is written slightly differently in ANSI C.

```
#include <stdio.h>
void main()
{
printf("Hello World n");
}
```

```
main()
{
printf("Hello World\n");
}
```

C includes several pre-defined data types including integers and floating point numbers.

- int (integer)
- long (long integer)
- float (floating point real)
- double (double precision real)

C allows more complex data types to be created using structs (similar to records in Pascal). The language allows the use of pointers to access memory locations, and the use of pointers provides flexibility. This allows the memory locations to be directly referenced and modified. For example:

```
int x;
int *ptr_x;
x = 4;
ptr_x = & x;
*ptr_x =5;
```

The address (or location in memory) of the variable x is given by &x. The variable ptr_x is then assigned to the address of x. C allows the content referenced by the pointer to be modified: i.e., *ptr_x = 5 assigns 5 to the memory location referenced by ptr_x. This has the effect of assigning 5 to x.

C is a block structured language, and a program is structured into functions (or blocks). Each function block contains its own variables and functions. A functions may call itself (i.e., recursion is allowed).

C has been criticized on a number of grounds. One key criticism of the language is that it is very easy to make errors in C programs, and to thereby produce undesirable results. For example, one of the easiest mistakes to make is to accidently write the assignment operator (=) for the equality operator (==). This totally changes the meaning of the original statement as can be seen below:

```
if (a == b)

a++;                    .... Program fragment A

else

a--

if (a = b)
```

```
a++;                    .... Program fragment B

else

a--
```

Both program fragments are syntactically correct. Program fragment A is what was intended. However, program fragment B is what was written and it has a totally different meaning from that intended. The philosophy of C to allow statements to be written as concisely as possible, and this is potentially dangerous.[13] The use of pointers potentially leads to problems (as uninitialised pointers can point anywhere in memory, and may therefore write anywhere in memory. This may lead to unpredictable results). Therefore, the effective use of C requires experienced programmers, well documented source code, and formal peer reviews of the source code by other developers.

3.4 Object-Oriented Languages

Object-oriented programming is a paradigm shift from the traditional way of programming. The traditional view of programming is that a program is a collection of functions, or a list of instructions to be performed on the computer. Object-oriented programming considers a computer program to be a collection of objects that act on each other. Each object is capable of receiving messages, processing data, and sending messages to other objects. That is, each object may be viewed as an independent entity or actor with a distinct role or responsibility.

An object is a "black box" which sends and receives *messages*. A black box consists of *code* (computer instructions) and *data* (information which these instructions operate on). The traditional way of programming kept code and data separate. For example, functions and data structures in the C programming language are not connected. A function in C can operate on many different types of structures, and a particular structure may be operated on by more than one function. However, in the object-oriented world code and data are merged into a single indivisible thing called an object.

The reason that an object is called a black box is that the user of an object never needs to look inside the box, since all communication to it is done via messages. Messages define the *interface* to the object. Everything an object can do is represented by its message interface. Therefore, there is no need to know anything about what is in the black box (or object) in order to use it. The approach to access to an

[13] It is very easy to write incomprehensible code in C and even a 1-line of C code can be incomprehensible. The maintenance of poorly written code is a challenge unless programmers follow good programming practice. This discipline needs to be enforced by formal reviews of the source code.

object only through its messages, while keeping the internal details private is called *information hiding*[14] and this dates back to work done by Parnas in the early 1970s.

The origins of object-oriented programming go back to the late 1960s with the invention of the Simula 67 language at the Norwegian Computing Research Centre[15] in Oslo. It introduced the notion of classes and also the concept of instances of classes.[16] Simula 67 influenced other languages, and this led to the development of the Smalltalk object-oriented language at Xerox in the mid-1970s. Xerox introduced the term *Object-oriented programming* for the use of objects and messages as the basis for computation. Many modern programming languages (e.g., Java and C++) support object-oriented programming. The main features of object-oriented languages are given in Table 3.1

Object-oriented programming has become very popular in large-scale software engineering, and it became the dominant paradigm in programming from the late 1980s. The proponents of object-oriented programming argue that it is easier to learn, and simpler to develop and maintain. Its growth in popularity was helped by the rise in popularity of Graphical User Interfaces (GUI), as the development of GUIs are well-suited to object-oriented programming. The C++ language has become very popular and it is an object-oriented extension of the C programming language.

Object-oriented features have been added to many existing languages including Ada, BASIC, Lisp, Fortran, and Pascal. However, adding object-oriented features to procedural languages that were not initially designed for object-oriented methods often led to problems with compatibility and maintainability of code. Smalltalk was developed by Xerox in the 1970s and it influenced later languages such as C++ and Java.

3.4.1 C++ and Java

Bjarne Stroustroup developed extensions to the C programming language to include classes, and this led to C++ which was released in 1983. C++ was designed to use the power of object-oriented programming, and to maintain the speed and portability of C. It is essentially a superset of C, and it provides a significant extension of C's capabilities. C++ does not force the programmer to use the object-oriented features of the language, and it is valid to use C++ as a procedural language. It is also used as a teaching language.

---

[14] Information hiding is a key contribution by Parnas to computer science. He has also done work on mathematical approaches to software inspections using tabular expressions [ORg:06], but this work is mainly of academic interest.

[15] The inventors of Simula-67 were Ole-Johan Dahl and Kristen Nygaard.

[16] Dahl and Nygaard were working on ship simulations and were attempting to address the huge number of combinations of different attributes from different types of ships. Their insight was to group the different types of ships into different classes of objects, with each class of objects being responsible for defining its own data and behaviour.

Table 3.1 Object-oriented paradigm

Feature	Description
Class	A class defines the abstract characteristics of a thing, including its attributes (or properties), and its behaviours (or methods). For example, the class of Cats defines all possible cats by listing the characteristics that feline creatures possess (e.g., breed, fur, colour and the ability to meow). The members of a class are termed objects.
Object	An object is a particular instance of a class with its own set of attributes. For example, the object "Mimi" is an instance of the class Cat: i.e., it is one particular cat with its own particular set of characteristics. Mimi may contain black fur and has her own distinct meow. The set of values of the attributes of a particular object is called its state.
Method	The methods associated with a class represent the behaviours of the objects in the class. These are things that an object does and represent how an object is used. The method is the action that is carried out, i.e., the code that gets executed when the message is sent to a particular object. For example, one behaviour of the object "Mimi" in the class cats is the ability to meow and this is one of Mimi's methods.
Message Passing	Message passing is the process by which an object sends data to another object, or asks the other object to invoke a method.
Inheritence	A class may have sub-classes (or children classes) that are more specialised versions of the class. A subclass inherits the attributes and methods of the parent class. Inheritance allows the programmer to create new classes from existing classes. The derived classes inherit the methods and data structures of the parent class. New methods may be added, or existing methods overridden to make the child class more specific. For example, the class Cat has several sub-classes such as the Egyptian Mau, the Manx, the Somali, and the Turkish Angora. Mimi is a member of the Egyptian Mau sub-class. The sub-classes of a class inherit attributes and behaviors from their parent classes, and may also introduce their own attributes. Inheritance is an "is–a" relationship: Mimi is an Egyptian Mau and an Egyptian Mau is a Cat, therefore Mimi inherits from both Egyptian Maus and Cats. Inheritence may be single or multiple, although multiple inheritence is not always supported.
Encapsulation (Information Hiding)	One of the fundamental principles of the object-oriented world is encapsulation. This principle is that the internals of an object are kept private to the object, and may not be accessed from outside the object. That is, encapsulation hides the details of how a particular class works. Knowledge of the engine of a car is not required in order to be able to drive the car. Similarly, programmers do not need to know all details of an object to be able to use it. Encapsulation requires a clearly specified interface around the services provided.
Abstraction	Abstraction simplifies complexity by modeling classes and removing all un-necessary detail. All essential detail is represented, and non-essential information is ignored.
Polymorphism	Polymorphism is behavior that varies depending on the class in which the behavior is invoked. Two or more classes may react differently to the same message.

Table 3.1 (continued)

Feature	Description
Polymorphism	The same name is given to methods in different subclasses. The individual methods may implement similar tasks but are differentiated by the type of arguments passed to them. The concept of polymorphism is often expressed by the phrase "one interface, multiple methods". For example, the Class of Animals has subclasses Cow and Cat. If Cow is commanded to speak this may result in a Moo, whereas if a Cat is commanded to speak this may result in a Meow. That is, a different response is given depending on the object to which it is applied.

One key difference between C++ and C is the concept of a class. A *class* was explained in Table 3.1 and it is an extension to the C concept of a structure. The main difference is that a C data structure can hold only data, whereas a C++ class may hold both data and functions. An *object* is an instantiation of a class: i.e., the class is essentially the type, whereas the object is essentially the variable of that type. Classes are defined in C++ by using the keyword class as follows:

```
class class_name
{
     access_specifier_1:
     member1;
     access_specifier_2:
     member2;
...
}
```

The members may be either data or function declarations, and an access specifier is included to specify the access rights for each member (e.g., private, public or protected). Private members of a class are accessible only by other members of the same class;[17] public members are accessible from anywhere where the object is visible; protected are accessible by other members of same class and also from members of their derived classes. This is illustrated in the example of the definition of the class rectangle:

```
class CRectangle
{
int x, y;
         public:
         void set_values (int,int);
         int area (void);
} rect;
```

[17] It is also possible to define friendship among classes and private members are also accessible by friends.

Java is an object-oriented programming language developed by James Gosling and others at Sun Microsystems in the early 1990s. The syntax of the language was influenced by C and C++. The language was designed with portability in mind, and the objective is to allow a program to be written once and executed anywhere. Platform independence is achieved by compiling the Java code into Java bytecode. The latter is simplified machine instructions specific to the Java platform.

This code is then run on a Java Virtual Machine (JVM) that interprets and executes the Java bytecode. The JVM is specific to the native code on the host hardware. The problem with interpreting bytecode is that it is slow compared to traditional compilation. However, Java has a number of techniques to address this including just in time compilation and dynamic recompilation. Java also provides automatic garbage collection. This is a very useful feature as it protects programmers who forget to deallocate memory (thereby causing memory leaks).

Java is a proprietary standard that is controlled through the Java Community Process. Sun Microsystems makes most of its Java implementations available without charge. The following is an example of the Hello World program written in Java.

```
class HelloWorld
{
        public static void main (String args[])
        {
        System.out.println ("Hello World!");
        }
}
```

3.5 Functional Programming Languages

Functional programming is quite distinct from imperative programming in that computation for functional programs involves the evaluation of mathematical functions. Imperative programming, in contrast, involves the execution of sequential (or iterative) commands that change the system state. The assignment statement alters the value of a variable. That is, the variable x may represent 2 at one point in program execution, and following the Pascal assignment statement x := 3 (i.e., x is assigned to 3), the value of x changes and is now 3. In other words, for an imperative program, the value of a variable x may change during program execution.

There are no side effects or changes of state for functional programs. The fact that the value of x will always be the same makes it easier to reason about functional programs than imperative programs, as the latter contain side affects. Functional programming languages provide referential transparency: i.e., equals may be substituted for equals, and if two expressions have equal values, then one can be substituted for the other in any larger expression without affecting the result of the computation. This makes it easier to reason about functional programs.

Functional programming languages use higher-order functions,[18] recursion, lazy and eager evaluation, monads,[19] and Hindley-Milner type inference systems.[20] These languages have mainly been used in academia, but there has been some industrial use, including the use of Erlang for concurrent applications in industry. The roots of functional programming languages are in the Lambda calculus developed by Church in the 1930s. Lambda calculus provides an abstract framework for describing mathematical functions and their evaluation. Church employed lambda calculus to prove that there is no solution to the decision problem for first order arithmetic in 1936. That is, he proved that there is no general algorithm which decides whether a given statement in first order arithmetic is true or false.

Any computable function can be expressed and evaluated using lambda calculus or Turing machines. The question of whether two expressions in the lambda calculus are equivalent cannot be solved by a general algorithm, and this result was proved prior to the decidability problem. Lambda calculus uses transformation rules, and one of these rules is variable substitution. The original calculus developed by Church was untyped; however, typed lambda calculi have been developed in recent years. Lambda calculus has influenced functional programming languages such as Lisp, ML and Haskell.

Functional programming uses the notion of higher-order functions. Higher-order take other functions as arguments, and may return functions as results. The derivative function ${}^{d}/_{dx} f(x) = f'(x)$ is a higher order function. It takes a function as an argument and returns a function as a result. The derivative of the function Sin(x) is given by Cos(x). Higher-order functions allow currying which is a technique developed by Schönfinkel. It allows a function with several arguments to be applied to each of its arguments one at a time, with each application returning a new (higher-order) function that accepts the next argument.

John McCarthy developed LISP at MIT in the late 1950s, and this language includes many of the features found in modern functional programming languages.[21] The functional programming language Scheme built upon the ideas in LISP, and simplified and improved upon the language.

Kenneth Iverson developed APL[22] in the early 1960s and this language influenced Backus's FP programming language. Backus described how functional programs could be built up following the principle of compositionality.

The ML programming language was created by Robin Milner at the University of Edinburgh in the early 1970s. David Turner developed the language Miranda at

---

[18] Higher order functions are functions that take functions as arguments or return a function as a results. They are known as operators (or functionals) in mathematics, and one example is the derivative function ${}^{dy}/_{dx}$ that takes a function as an argument and returns a function as a result.

[19] Monads are used in functional programming to express input and output operations without introducing side effects. The Haskell functional programming language makes use of this feature.

[20] This is the most common algorithm used to perform type inference. Type inference is concerned with determining the type of the value derived from the eventual evaluation of an expression.

[21] Lisp is a multi-paradigm language rather than a functional programming language.

[22] Iverson received the Turing Award in 1979 for his contributions to programming language and mathematical notation. The title of his Turing award paper was "Notation as a tool of thought".

the University of Kent in the mid-1980s. The Haskell programming language was released in the late 1980s.

3.5.1 Miranda

Miranda was developed by David Turner at the University of Kent, England in the mid-1980s [Turn:85]. It is a non-strict functional programming language: i.e., the arguments to a function are not evaluated until they are actually required within the function being called. This is also known as lazy evaluation, and one of its main advantages is that it allows an infinite data structures to be passed as an argument to a function. Miranda is a pure functional language in that there are no side-effect features in the language. The language has been used for:

- Rapid prototyping
- Specification language
- Teaching Language

A Miranda program is a collection of equations that define various functions and data structures. It is a strongly typed language with a terse notation.

```
z = sqr p / sqr q
sqr k = k * k
p = a + b
q = a − b
a = 10
b = 5
```

The scope of a formal parameter (e.g., the parameter k above in the function sqr) is limited to the definition of the function in which it occurs.

One of the most common data structures used in Miranda is the list. The empty list is denoted by [ ], and an example of a list of integers is given by [1, 3, 4, 8]. Lists may be appended to by using the "++" operator. For example:

[1, 3, 5] ++ [2, 4] is [1, 3, 5, 2, 4].

The length of a list is given by the "#" operator:

[1, 3] = 2

The infix operator ":" is employed to prefix an element to the front of a list. For example:

5 : [2, 4, 6] is equal to [5, 2, 4, 6]

The subscript operator "!" is employed for subscripting: For example:

Nums = [5, 2, 4, 6] then Nums!0 is 5.

The elements of a list are required to be of the same type. A sequence of elements that contains mixed types is called a tuple. A tuple is written as follows:

Employee = ("Holmes", "222 Baker St. London", 50, "Detective")

A tuple is similar to a record in Pascal whereas lists are similar to arrays. Tuples cannot be subscripted but their elements may be extracted by pattern matching. Pattern matching is illustrated by the well-known example of the factorial function:

```
fac 0 = 1
fac (n+1) = (n+1) * fac n
```

The definition of the factorial function uses two equations, distinguished by the use of different patterns in the formal parameters. Another example of pattern matching is the reverse function on lists:

```
          reverse [] = []
reverse (a:x) = reverse x ++ [a]
```

Miranda is a higher-order language, and it allows functions to be passed as parameters and returned as results. Currying is allowed and this allows a function of n-arguments to be treated as n applications of a function with 1-argument. Function application is left associative: i.e., f x y means (f x) y. That is, the result of applying the function f to x is a function, and this function is then applied to y. Every function of two or more arguments in Miranda is a higher-order function.

3.5.2 Lambda Calculus

Lambda Calculus (λ-calculus) was designed by Alonzo Church in the 1930s to study computability. It is a formal system that may be used to study function definition, function application, parameter passing and recursion. Lambda calculus may be employed to define what a computable function is, and any computable function may be expressed and evaluated using the calculus. Church used lambda calculus in 1936 to give a negative answer to Hilbert's Entscheidungs problem. The question of whether two lambda calculus expressions are equivalent cannot be solved by a general algorithm.

The lambda calculus is equivalent to the Turing machine formalism. However, lambda calculus emphasises the use of transformation rules, whereas Turing machines are concerned with computability on primitive machines. Lambda calculus consists of a small set of rules:

- Alpha-conversion rule (α-conversion)[23]
- Beta-reduction rule (β-reduction)[24]
- Eta-conversion (η-conversion)[25]

Every expression in the λ-calculus stands for a function with a single argument. The argument of the function is itself a function with a single argument, and so on. The definition of a function is anonymous in the calculus. For example, the function that adds one to its argument is usually defined as $f(x) = x + 1$. However, in λ-calculus the function is defined as:

$\lambda\, x.x + 1$ (or equivalently as $\lambda\, z.z + 1$)

The name of the formal argument x is irrelevant and an equivalent definition of the function is $\lambda\, z.\, z+1$. The evaluation of a function f with respect to an argument (e.g. 3) is usually expressed by $f(3)$. In λ-calculus this would be written as $(\lambda\, x.x + 1)3$, and this evaluates to $3+1 = 4$. Function application is left associative: i.e., $f\ x\ y = (f\ x)\ y$. A function of two variables is expressed in lambda calculus as a function of one argument which returns a function of one argument. This is known as currying and has been discussed earlier. For example, the function $f(x, y) = x + y$ is written as $\lambda\, x.\, \lambda\, y.\, x + y$. This is often abbreviated to $\lambda\, x\ y.\, x + y$.

λ-calculus is a simple mathematical system and its syntax is defined as follows:

<exp> ::= <identifier> |
λ <identifier>.<exp> | --abstraction
<exp> <exp> | --application
(<exp>)

λ-Calculus's four lines of syntax plus *conversion* rules are sufficient to define Booleans, integers, data structures and computations on them. It inspired Lisp and modern functional programming languages.

3.6 Logic Programming Languages

Imperative programming languages such as Fortran, C, and Pascal require the programmer to explicitly define an algorithm to solve the particular problem, and the

---

[23] This essentially expresses that the names of bound variables is unimportant.
[24] This essentially expresses the idea of function application.
[25] This essentially expresses the idea that two functions are equal if and only if they give the same results for all arguments.

program then executes the algorithm. Imperative languages are concerned with implementation details on how the problem will be solved. Logic programming languages, in contrast, describe what is to be done, rather than how it should be done. These languages are concerned with the statement of the problem to be solved, rather than how the problem will be solved.

These programming languages use mathematical logic as a tool in the statement of the problem definition. Logic is a useful tool in developing a body of knowledge (or theory), and it allows rigorous mathematical deduction to derive further truths from the existing set of truths. The theory is built up from a small set of axioms or postulates and rules of inference are employed to derive further truths logically. The objective of logic programming is to employ mathematical logic to assist with computer programming. Many problems are naturally expressed as a theory, and the statement that a problem needs to be solved is often equivalent to determining if a new hypothesis is consistent with an existing theory. Logic provides a rigorous way to determine this, as it includes a rigorous process for conducting proof.

Computation in logic programming is essentially logical deduction, and logic programming languages use first-order[26] predicate calculus. These languages employ theorem proving to derive a desired truth from an initial set of axioms. These proofs are constructive[27] in that more than existence is demonstrated: in fact, an actual object that satisfies the constraints is produced. Logic programming specifies the objects, the relationships between them and the constraints that must be satisfied for the problem.

1. The set of objects involved in the computation
2. The relationships that hold between the objects
3. The constraints for the particular problem.

The language interpreter decides how to satisfy the particular constraints. Artificial Intelligence influenced the development of logic programming, and John McCarthy[28] demonstrated that mathematical logic could be used for expressing knowledge. The first logic programming language was Planner, and this language was designed by Carl Hewitt at MIT in 1969. Planner uses a procedural approach for knowledge representation rather than McCarthy's approach of expressing knowledge using mathematical logic. The language allows theorem proving.

[26] First-order logic allows quantification over objects but not functions or relations. Higher-order logics allow quantification of functions and relations.

[27] For example, the statement $\exists x$ such that $x = \sqrt{4}$ states that there is an x such that x is the square root of 4, and the constructive existence yields that the answer is that $x = 2$ or $x = -2$ i.e., constructive existence provides more the truth of the statement of existence, and an actual object satisfying the existence criteria is explicitly produced.

[28] John McCarthy received the Turing Award in 1971 for his contributions to Artificial Intelligence. He also developed the programming language LISP.

The best known logic programming languages is Prolog, and this language was developed in the early 1970s by Alain Colmerauer and Robert Kowalski. It stands for *pro*gramming in *log*ic. It is a goal-oriented language that is based on predicate logic. Prolog became an ISO standard in 1995. The language attempts to solve a goal by tackling the sub-goals that the goal consists of:

```
goal :- subgoal1, ..., subgoaln.
```

which states that in order to prove a particular goal, it is sufficient to prove $subgoal_1$ through $subgoal_n$. Each line of a typical Prolog program consists of a rule or a fact, and the language specifies what exists rather than how. The following program fragment has one rule and two facts:

```
grandmother(G,S) :- parent(P,S), mother(G,P).
mother(sarah, issac).
parent(issac, jacob).
```

The first line in the program fragment is a rule that states that G is the grandmother of S if there is a parent P of S and G is the mother of P. The next two statements are facts stating that issac is a parent of jacob, and that sarah is the mother of issac. A particular goal clause is true if all of its subclauses are true:

```
goalclause(Vg):- clause1(V1),..,clausem(Vm)
```

A Horn clause consists of a goal clause and a set of clauses that must be proven separately. Prolog finds solutions by *unification*: i.e., by binding a variable to a value. For an implication to succeed, all goal variables V*g* on the left side of :- must find a solution by binding variables from the clauses which are activated on the right side. When all clauses are examined and all variables in V*g* are bound, the goal succeeds. But if a variable can not be bound for a given clause, that clause fails. Following the failure of a fails, Prolog *backtracks*. This involves going back to the left to a previous clauses to continue trying to unify with alternative bindings. Backtracking give Prolog the ability to find multiple solutions to for a given query or goal.

Most logic programming languages use a simple searching strategy to consider alternatives:

- If a goal succeeds and there are more goals to achieve, then remember any untried alternatives and go on to the next goal.
- If a goal is achieved and there are no more goals to achieve then stop with success.
- If a goal fails and there are alternative ways to solve it then try the next one.
- If a goal fails and there are no alternate ways to solve it, and there is a previous goal, then go back to the previous goal.
- If a goal fails and there are no alternate ways to solve it, and no previous goal, then stop with failure.

Constraint programming is a programming paradigm where relations between variables can be stated in the form of constraints. Constraints specify the properties of the solution, and differs from the imperative programming languages in that they do not specify the sequence of steps to execute.

3.7 Syntax and Semantics

There are two key parts to any programming language and these are its syntax and semantics. The syntax is the grammar of the language, and a program needs to be syntactically correct with respect to its grammar. The semantics of the language is deeper, and determines the meaning of what has been written by the programmer. The semantics of a language determines what a syntactically valid program will compute. A programming language is therefore given by:

Programming Language = Syntax + Semantics

The theory of the syntax of programming languages is well established, and Backus Naur Form[29] (BNF) is employed to specify the grammar of languages. The grammar of a language may be input into a parsers, and the parser is then employed to determine if a program is syntactically correct. BNF was employed to define the grammar for the Algol-60 programming language, and is widely used today to specify the grammars of computer programming languages. Chomsky[30] identified a number of different types of grammar (regular, context free, context sensitive). A BNF specification consists of a set of rules such as:

```
<symbol> ::= <expression with symbols>
```

where <symbol> is a *nonterminal*, and the expression consists of sequences of symbols and/or sequences separated by the vertical bar "|" which indicates a choice. That is, it is one possible substitution for the symbol on the left. Symbols that never appear on a left side are called *terminals*.

The definition of the syntax of various statements in a programming language is given below:

```
<loop statement> ::= <while loop>|<for loop>
<while loop> ::= while ( <condition> ) <statement>
<for loop> ::= for ( <expression> ) <statement>
<statement> ::= <assignment statement>|<loop statement>
<assignment statement> ::= <variable> := <expression>
```

---

[29] Backus Naur Form is named after John Backus and Peter Naur. It was created as part of the design of Algol 60, and used to define the syntax rules of the language.

[30] Chomsky made important contributions to linguistics and the theory of grammars. He is more widely known today as a critic of United States foreign policy.

The example above is a partial definition of the syntax of various statements in the programming language. It includes various non-terminals (<loop statement>, <while loop>, <for loop>, <condition>, <expression>, <statement>, <assignment statement>, and <variable>). The terminals include "while", "for", ":=", "(" and ")". The production rules for <condition> and <expression> are not included.

There are various types of grammars such as regular grammars, context free grammars, and context sensitive grammars. The grammar of a language (e.g. LL(1), LL(k), LR(1), LR(k) grammar expressed in BNF notation) is translated by a parser into a parse table. Each type of grammar has its own parsing algorithm to determine whether a particular program is valid with respect to its grammar.

3.7.1 Programming Language Semantics

The formal semantics of a programming language is concerned with the meaning of programs. A programmer writes a program according to the rules of the programming language. The compiler first checks the program for syntactic correctness: i.e., it determines whether the program as written may be generated from the grammar of the programming language. If the program is syntactically correct, then the compiler generates machine code that corresponds to what the programmer has written.[31]

The compiler must preserve the semantics of the language: i.e., a program's syntax gives no information on the meaning of the program: this is given by the semantics of the programming language, and it is the role of the compiler to preserved the semantics of the language. In natural languages, it is possible for sentences to be syntactically correct, but semantically meaningless.[32] Similarly, in programming languages it is possible to write syntactically correct programs that behave in quite a different way from the intentions of the programmer.

The formal semantics of a language is given by a mathematical model that describes the possible computations described by the language. There are three main approaches to programming language semantic, and these are axiomatic semantics, operational semantics and denotational semantics (Table 3.2).

---

[31] Of course, what the programmer has written may not be what the programmer had intended.

[32] An example of a sentence in a natural language that is syntactically correct but semantically meaningless is "I will be here yesterday".

Table 3.2 Programming language semantics

Approach	Description
Axiomatic Semantics	Axiomatic semantics involves giving meaning to phrases of the language by describing the logical axioms that that apply to them. This is an approach that is based on mathematical logic, and employs pre and post condition assertions to specify what happens when the statement executes. The relationship between the initial assertion and the final assertion essentially gives the semantics of the code. This approach is due to C.A.R. Hoare[33] and appeared in his famous 1969 paper [Hor:69] "An axiomatic basis for computer programming". The axiomatic approach is often used in proving program correctness. Another use of axiomatic semantics has been in using the assertions as program specification from which the program code may be derived (to satisfy the specification). Dijkstra [Dij:76] has argued for developing the program and its proof of correctness together. The basic form of axiomatic semantics is {P} S {Q}, and this states that if P is true before S is executed, then Q is true after S. Axiomatic semantics does not employ the concept of the state of the machine in giving meaning to programs.
Operational Semantics	The operational semantics for a programming language describes how a valid program is interpreted as sequences of computational steps. These sequences then define the meaning of the program. Operational semantics was developed by Gordon Plotkin [Plo:81], and is essentially a mathematical interpreter. However, operational semantics is more precise than an interpreter since it is defined mathematically, and not based on the meaning of the language in which the interpreter is written. An abstract machine (SECD machine) may be defined to give meaning to phrases, and this is done by describing the transitions they induce on states of the machine. Operational semantics may also be defined using the lambda calculus.
Denotational Semantics	Denotational semantics (originally called mathematical semantics) provides meaning to programs in terms of mathematical objects such as integers, tuples and functions. It was developed by Christopher Strachey and Dana Scott at the Programming Research Group at Oxford, England in the mid-1960s, and their approach to semantics is known as the Scott-Strachey approach. Dana Scott's contributions included the formulation of domain theory, and this allowed programs containing recursive functions and loops to be given a precise semantics. Each phrase in the language is translated into a mathematical object that is the *denotation* of the phrase. Denotational Semantics has been applied to language design and implementation.

3.8 Review Questions

1. Describe the five generations of programming languages.
2. Describe the early use of machine code including its advantages and disadvantages.

3. Describe the early use of assembly languages including their advantages and disadvantages.
4. Describe the key features of Fortran and Cobol.
5. Describe the key features of Pascal and C.
6. Discuss the key features of object-oriented languages.
7. Discuss the similarities and differences between imperative programming languages and functional programming languages.
8. Discuss the key features of logic programming languages.
9. Discuss the similarities and differences between syntax and semantics of programming languages.
10. Discuss the three main types of programming language semantics.

3.9 Summary

This chapter considered the evolution of programming languages from the older machine languages, to the low-level assembly languages, to high-level programming languages and object-oriented languages, to functional and logic programming languages. Finally, the syntax and semantics of programming languages was considered.

The advantages of the machine-level programming languages were execution speed and efficiency. It was difficult to write programs in these languages as the statements in the language are just a stream of binary numbers. Further, these languages were not portable, as the machine language for one computer could differ significantly from the machine language of another.

The second generation languages, or 2GL, are low-level assembly languages that are specific to a particular computer and processor. These are easier to write and understand. They must be converted into the actual machine code to run on the computer. The assembly language is specific to a particular processor family and environment, and is therefore not portable. However, their advantages are execution speed, as the assembly language is the native language of the processor, and compilation.

The third generation languages, or 3GL, are high-level programming languages. They are general purpose languages and have been applied to business, scientific and general applications. They are designed to be easier for a human to understand and to allow the programmer to focus on problem solving. Their advantages include ease of readability, portability, and ease of debugging and maintenance. The early 3GLs were procedure-oriented and the later 3GLs were object-oriented.

Fourth-generation languages, or 4GLs, are languages that consist of statements similar to human language. Most fourth generation languages are non-procedural,

[33] Hoare was influenced by earlier work by Floyd on assigning meanings to programs using flowcharts [Flo:67].

and are often used in database programming. They specify what needs to be done rather than how it should be done.

Fifth-generation programming languages or 5GLs, are programming languages that is based around solving problems using logic programming or applying constraints to the program. They are designed to make the computer (rather than the programmer) solve the problem. The programmer only needs to be concerned with the specification of the problem and the constraints to be satisfied, and does not need to be concerned with the algorithm or implementation details. These languages are used mainly in academic environments.

Chapter 4
Software Engineering

Key Topics

Birth of Software Engineering
- Floyd
- Hoare
- VDM and Z
- Cleanroom and Software Reliability
- Software Inspections and Testing
- Project Management
- Software Process Maturity Models

4.1 Introduction

The NATO Science Committee organized two famous conferences on software engineering in the late 1960s. The first conference was held in Garmisch, Germany, in 1968 and this was followed by a second conference in Rome in 1969. The Garmisch conference was attended by over 50 people from 11 countries including the eminent Dutch computer scientist, Edger Djkstra (Fig. 4.1).

The NATO conferences highlighted the problems that existed in the software sector in the late 1960s, and the term *software crisis* was coined to refer to the problems associated with software projects. These included budget and schedule overruns, and problems with the quality and reliability of the delivered software. This led to the birth of *software engineering* as a separate discipline, and the realization that programming is quite distinct from science and mathematics. Programmers are like engineers in the sense that they build products; however, programmers are not educated as traditional engineers as they receive minimal education in design and mathematics.[1]

[1] Software Companies that are following approaches such as the CMM or ISO 9000:2000 consider the qualification of staff before assigning staff to performing specific tasks. The approach adopted

Fig. 4.1 Edsger Dijkstra at NATO Conference Courtesy of Brian Randell.

The construction of bridges was problematic in the nineteenth century, and many people who presented themselves as qualified to design and construct bridges did not have the required knowledge and expertise. Consequently, many bridges collapsed, endangering the lives of the public. This led to legislation requiring an engineer to be licensed by the Professional Engineering Association prior to practicing as an engineer. These engineering associations identify a core body of knowledge that the engineer is required to possess, and the licensing body verifies that the engineer has the required qualifications and experience. The licensing of engineers by most branches of engineering ensures that only personnel competent to design and build

is that the appropriate qualifications and experience for the role are considered prior to appointing a person to carry out a particular role. My experience is that the more mature companies place significant emphasis on the education and continuous development of their staff, and on introducing best practice in software engineering into their organization. I have observed an increasing trend among companies to mature their software processes to enable them to deliver superiour results. One of the purposes that the original CMM served was to enable the United States Department of Defence (DOD) to have a mechanism to assess the capability and maturity of software subcontractors.

products actually do so. This in turn leads to products that the public can safely use. In other words, the engineer has a responsibility to ensure that the products are properly built, and are safe for the public to use.

Parnas argues that traditional engineering be contrasted with the software engineering discipline where there is no licensing mechanism, and where individuals with no qualifications can participate in building software products.[2] However, the fact that the maturity frameworks such as the CMMI or ISO 9000 place a strong emphasis on qualifications and training may help to deal with this.

The Standish group conducted research in the late 1990s [Std:99] on the extent of current problems with schedule and budget overruns of IT projects. This study was conducted in the United States, but there is no reason to believe that European or Asian companies perform any better. The results indicate serious problems with on-time delivery.[3] Fred Brooks argues that software is inherently complex, and that there is no silver bullet that will resolve all of the problems associated with software such as schedule overruns and software quality problems [Brk:75, Brk:86] (Fig. 4.2).

The problem with poor software quality and poor software design is evident in the security flaws exhibited in early versions of Microsoft Windows software. Such flaws can cause minor irritation at best or at worse can seriously disrupt the work of an organization or individual. The Y2K problem, where dates were represented in a 2-digit format, required major rework for year 2000 compliance. Clearly, well-designed programs would have hidden the representation of the date, and this would have required minimal changes for year 2000 compliance. However, the quality of software produced by some companies is superiour.[4] These companies employ mature software processes and are committed to continuous improvement.

Mathematics plays a key role in engineering and may potentially assist software engineers in the delivery of high-quality software products that are safe to use. Several mathematical approaches that can assist in delivering high-quality software are described in [ORg:06]. There is a lot of industrial interest in software process maturity for software organizations, and approaches to assess and mature software companies are described in [ORg:02].[5] These focus mainly on improving

---

[2] Modern HR recruitment specifies the requirements for a particular role, and interviews with candidates aim to establish that the candidate has the right education and experience for the role.

[3] It should be noted that these are IT projects covering diverse sectors including banking, telecommunications, etc., rather than pure software companies. My experience is that mature software companies following maturity frameworks such as level 2 or level 3 CMM maturity achieve project delivery generally within 20% of the project estimate. Mathematical approaches to software quality [ORg:06] are focused on technical ways to achieve software quality. There is also the need to focus on the management side of software engineering as this is essential for project success. The reader is referred to [ORg:02].

[4] I recall projects at Motorola that regularly achieved 5.6 σ-quality in a L4 CMM environment (i.e., approx 20 defects per million lines of code. This represents very high quality).

[5] Approaches such as the CMM or SPICE (ISO 15504) focus mainly on the management and organizational practices required in software engineering. The emphasis is on defining and following the software process. In practice, there is often insufficient technical detail on requirements,

Fig. 4.2 Fred Brooks Photo by Dan Sears.

the effectiveness of the management, engineering and organization practices related to software engineering.

The use of mathematical techniques is complementary to a maturity approach such as the Capability Maturity Model (CMM). The CMM allows the mathematical techniques to be properly piloted to verify that the approach suits the company, and that superior results will be achieved by using these methods. The bottom line for a software company is that the use of these approaches should provide a cost-effective solution. The CMM also includes practices to assist with a sound deployment of a new process.

4.2 What Is Software Engineering?

Software engineering involves multi-person construction of multi-version programs. The IEEE 610.12 definition of Software Engineering is:

design, coding and testing in the models, as the models focus on what needs to be done rather how it should be done.

Definition 4.1 (Software Engineering) Software engineering is the application of a systematic, disciplined, quantifiable approach to the development, operation, and maintenance of software; that is, the application of engineering to software, and the study of such approaches.

Software engineering includes:

1. Methodologies to determine requirements, design, develop, implement and test software to meet customers needs.
2. The philosophy of engineering: i.e., an engineering approach to developing software is adopted. That is, products are properly designed, developed, tested, with quality and safety properly addressed.
3. Mathematics may be employed to assist with the design and verification of software products. The level of mathematics to be employed will depend on the safety critical nature of the product, as systematic peer reviews and testing are often sufficient in building quality into the software product.
4. Sound project management and quality management practices are employed.

Software engineering requires the engineer to state precisely the requirements that the software product is to satisfy, and then to produce designs that will meet these requirements. Engineers should start with a precise description of the problem to be solved; they then proceed to producing a design and validate the correctness of the design; finally, the design is implemented and testing is performed to verify its correctness with respect to the requirements. The software requirements should clearly state what is required, and it should also be evident what is not required.

Engineers produce the product design, and then analyse their design for correctness. This analysis may include mathematics and software inspections, and this analysis is essential to ensure that the design is correct. The term *engineer* is generally applied only to people who have attained the necessary education and competence to be called engineers, and who base their practice on mathematical and scientific principles. Often in computer science the term engineer is employed loosely to refer to anyone who builds things, rather than to an individual with a core set of knowledge, experience, and competence.

Parnas[6] is a strong advocate of a classical engineering approach, and argues that computer scientists should have the right education to apply scientific and mathematical principles to their work. Software engineers need to receive an appropriate education in mathematics and design, to enable them to be able to build high-quality and safe products. He has argued that computer science courses often tend to emphasize the latest programming language rather than the essential engineering foundations. Therefore, the material learned in a computer science course is often outdated shortly after the student graduates. That is, the problem is that students are generally taught programming and syntax, but not how to design and analyse

[6] Parnas's key contribution to software engineering is information hiding which is used in the object-oriented world. He has also done work (mainly of academic interest) on mathematical approaches to software quality.

software. Computer science courses tend to include a small amount of mathematics whereas mathematics is a significant part of an engineering course. The engineering approach to the teaching of mathematics is to emphasize its application and especially the application to developing and analysing product designs. The mathematics that software engineering students need to be taught includes sets, relations, functions, mathematical logic, and finite state machines. The emphasis in engineering is on the application of mathematics to solve practical problems, rather than theoretical mathematics for its own sake.

The consequence of the existing gap in the education of current software engineers is that there is a need to retrain existing software engineers with a more solid engineering education.[7] Software engineers need education on specification, design, turning designs into programs, software inspections, and testing. The education should enable the software engineer to produce well-designed programs that will correctly implement the requirements.

Parnas argues that software engineers have individual responsibilities as professional engineers.[8] They are responsible for designing and implementing high-quality and reliable software that is safe to use. They are also accountable for their own decisions and actions,[9] and have a responsibility to object to decisions that violate professional standards. Professional engineers have a duty to their clients to ensure that they are solving the real problem of the client. They need to precisely state the problem before working on its solution. Engineers need to be honest about current capabilities, and when asked to work on problems that have no appropriate technical solution, this should be stated honestly rather than accepting a contract for something that cannot be done.

---

[7] Realistically, a mass re-education program for current software engineers is unrealistic and highly unlikely to take place. Any retraining program for current software engineers would need to minimize the impact on company time, as software companies in the new millennium are very focused on the bottom line, as they are working in a very competitive environment. This includes intense competition from the low cost software development locations in Eastern Europe and Asia as outsourcing of software development to such locations is an increasing trend.

[8] The concept of accountability is not new; indeed the ancient Babylonians employed a code of laws c. 1750 BC. known as the Hammarabi Code. This code included the law that if a house collapsed and killed the owner then the builder of the house would be executed.

[9] However, it is unlikely that an individual programmer would be subject to litigation in the case of a flaw in a program causing damage or loss of life. Most software products are accompanied by a comprehensive disclaimer of responsibility for problems rather than a guarantee of quality. Software engineering is a team-based activity involving many engineers in various parts of the project, and it could be potentially difficult for an outside party to prove that the cause of a particular problem is due to the professional negligence of a particular software engineer, as there are many others involved in the process such as reviewers of documentation and code and the various test groups. Companies are more likely to be subject to litigation, as a company is legally responsible for the actions of their employees in the workplace, and the fact that a company is a financially richer entity than one of its employees. However, the legal aspects of licensing software may protect software companies from litigation including those companies that seem to place little emphasis on software quality. However, greater legal protection for the customer can be built into the contract between the supplier and the customer for bespoke-software development.

The licensing of a professional engineer provides confidence that the engineer has the right education and experience to build safe and reliable products.[10] Otherwise, the profession gets a bad name as a result of poor work carried out by unqualified people. Professional engineers are required to follow rules of good practice, and to object when the rules are violated.[11] The licensing of an engineer requires that the engineer completes an accepted engineering course and understands their responsibilities. The professional body is responsible for enforcing standards and certification. The term "engineer" is a title that is awarded on merit, but it also places responsibilities on its holder.

Engineers have a professional responsibility and are required to behave ethically with their clients. The membership of the professional engineering body requires the member to adhere to the code of ethics of the profession. The code of ethics[12] will detail the ethical behaviour and responsibilities (Table 4.1).

The approach used in current software engineering is to follow a well-defined software engineering process. The process includes activities such as project management, requirements gathering, requirements specification, architecture design, software design, coding, and testing. Most companies use a set of templates for the various phases. The waterfall model [Roy:70] and spiral model [Boe:88] are popular software development lifecycles.

The waterfall model (Fig. 4.3) starts with requirements, followed by specification, design, implementation, and testing. It is typically used for projects where the requirements can be identified early in the project lifecycle or are known in advance. The waterfall model is also called the "V" life cycle model, with the left-hand side of the "V" detailing requirements, specification, design, and coding and the right-hand side detailing unit tests, integration tests, system tests and acceptance testing.

Table 4.1 Professional responsibilities of software engineers

No.	Responsibility
1.	Honesty and fairness in dealings with clients.
2.	Responsibility for actions.
3.	Continuous learning to ensure appropriate knowledge to serve the client effectively.

[10] Companies that are following the CMMI framework generally employ a highly rigorous process for supplier selection. The formal evaluation of the supplier determines the technical knowledge of the candidate suppliers as well as their software process maturity and capability. The evaluation will generally take into account the qualifications and experience of the technical team that will be carrying out the work.

[11] Software companies that are following the CMMI or ISO 9000 will employ audits to verify that the rules have been followed. Auditors report their findings to management and the findings are addressed appropriately by the project team and affected individuals.

[12] My experience of working in software companies is that these are core values of most mature software companies. Larger companies have a code of ethics that employees are required to adhere as larger corporate companies will wish to project a good corporate image and to be respected world-wide.

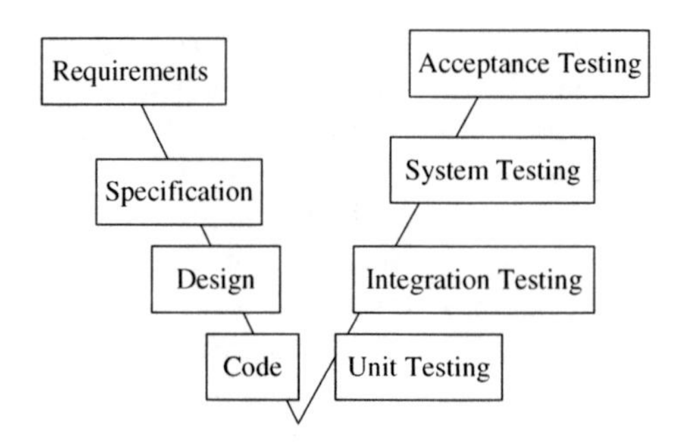

Fig. 4.3 Waterfall lifecycle model (V-Model)

Each phase has entry and exit criteria which must be satisfied before the next phase commences. There are many variations to the waterfall model.

The spiral model (Fig. 4.4) is another lifecycle model, and is useful where the requirements are not fully known at project initiation. There is an evolution of the requirements as a part of the development lifecycle. The development proceeds in a number of spirals, where each spiral typically involves updates to the requirements, design, code, testing, and a user review of the particular iteration or spiral.

The spiral is, in effect, a re-usable prototype and the customer examines the current iteration and provides 'feedback to the development team to be included in the next spiral. This approach is often used in joint application development for web-based software development. The approach is to partially implement the system. This leads to a better understanding of the requirements of the system and it then feeds into the next cycle in the spiral. The process repeats until the requirements and product are fully complete.

There are other life-cycle models, for example, the iterative development process which combines the waterfall and spiral lifecycle model. The cleanroom approach

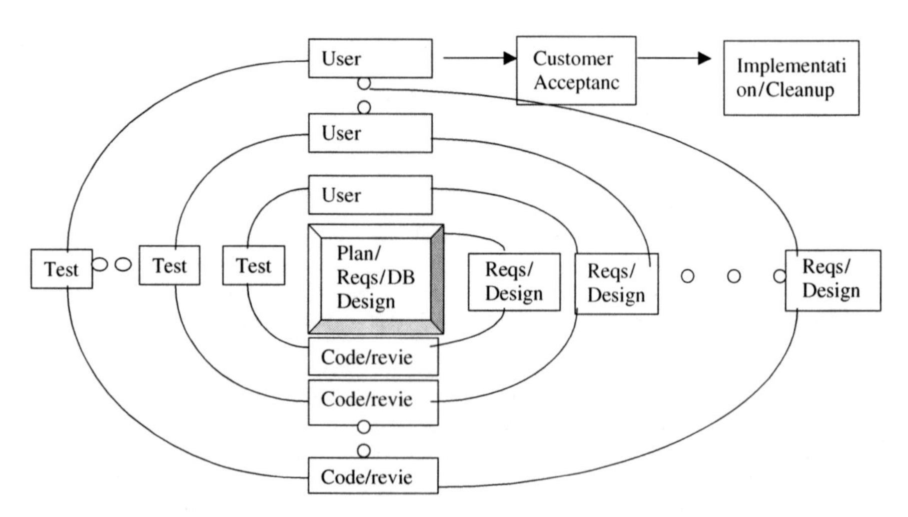

Fig. 4.4 Spiral lifecycle model

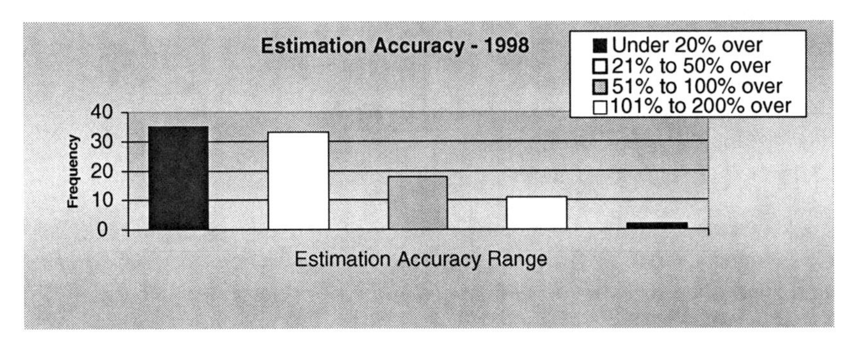

Fig. 4.5 Standish group report – estimation accuracy

to software development includes a phase for formal specification and its approach to testing is quite distinct from other models as it is based on the predicted usage of the software product. Finally, the Rational Unified Process (RUP) has become popular in recent years.

The challenge in software engineering is to deliver high-quality software on time to customers. The Standish Group research [ORg:02] (Fig. 4.5) on project cost overruns in the United States during 1998 indicate that 33% of projects are between 21% and 50% over estimate, 18% are between 51% and 100% over estimate, and 11% of projects are between 101% and 200% overestimate.

Project management and estimating project cost and schedule accurately are key software engineering challenges. Consequently, organizations need to determine how good their estimation process actually is and to make improvements as appropriate. The use of software metrics is one mechanism to determine the current effectiveness of software estimation. This involves computing the variance between actual project effort and estimated project estimate, and actual project schedule versus projected project schedule. Risk management is a key part of project management, and its objective is to identify potential risks to the project; determine the probability of the risks occurring; assessing the impact of each risk if it materializes; identifying actions to eliminate the risk or to reduce the probability of it occurring or its impact should it occur; contingency plans in place to address the risk if it materializes; and finally to track and manage the risks throughout the project.

The concept of process maturity has become popular with the Capability Maturity Model, and organizations such as the SEI have collected empirical data to suggest that there is a close relationship between software process maturity and the quality and the reliability of the delivered software. However, the main focus of the CMM is management and organization practices rather than on the technical engineering practices. However, there are software engineering practices at maturity level 3 in the model. The use of the CMMI does provide a good engineering approach, as it does place strict requirements on what processes a company needs to have in place to provide a good engineering solution. This includes:

- Developing and managing requirements
- Doing effective design
- Planning and tracking projects
- Building quality into the product with peer reviews
- Performing rigourous testing
- Performing independent audits

There has been a growth of popularity among software developers in light-weight methodologies such as XP [Bec:00]. These methodologies view documentation with distaste, and often software development commences prior to the full specification of the requirements.

Classical engineering places emphasis on detailed planning and design and includes appropriate documentation. The design documents are analyzed and reviewed and used as a reference during the construction. Documentation is produced in the software sector with varying degrees of success. The popularity of a methodology such as XP suggests developer antipathy to documentation. However, the more mature companies recognize the value of software documentation, and regard documentation as essential for all phases of the project. It is essential that the documentation is kept up to date and reflects the actual system since the impact of any changes requested to the software during maintenance cannot be properly determined. The empirical evidence suggests that software documentation is sometimes out of date.

The difference between traditional engineering documents and standard software documents (e.g., documentation following the IEEE templates) is that engineering documents are precise enough to allow systematic analysis to be carried out, whereas since software documents employ natural language rather than mathematics. Parnas argues that only limited technical evaluation may take place on standard software engineering documents. He argues that if software engineers are to perform as engineers then the software documents should be similar to engineering documents, and should include sufficient mathematics to allow rigorous analysis to be performed. However, formal software inspections by engineers generate good results on documents that have been written in natural language, and help in ensuring that the document is fit for purpose. The analysis of traditional engineering documents uses mathematics to verify that the design is correct.

4.3 Early Software Engineering

Robert Floyd and others did pioneering work on software engineering in the 1960s. Floyd made valuable contributions to the theory of parsing, the semantics of programming languages, program verification, and methodologies for the creation of efficient and reliable software. He was born in New York in 1936, and attended the university of Chicago. He became a computer operator in the early 1960s, and commenced publishing papers in computer science. He was appointed associate professor at Carnegie Mellow University in 1963 and became a full professor of computer science at Sanford University in 1969. Knuth remarked that although Floyd never actually obtained a Ph.D. that several of his papers were better than any Ph.D. thesis that he saw.

Floyd did pioneering work on mathematical techniques to verify program correctness. Mathematics and Computer Science were regarded as two completely separate disciplines in the 1960s, and the accepted approach to software development at that time was based on the assumption that the completed code would always contain defects. It was therefore better and more productive to write the code as quickly as possible, and to then perform various tests to find the defects. Programmers then corrected the defects, made patches and re-tested and found more defects. This continued until they could no longer find defects. Of course, there was always the danger that defects remained in the code that could give rise to software failures.

Floyd believed that there was a way to construct a rigorous proof of the correctness of the programs using mathematics, and that the disciplines of mathematics and computer science were related. Floyd showed that mathematics could be used for program verification, and he introduced the concept of assertions that provided programmers with a way to verify the correctness of their programs.

Flowcharts were employed in the 1960s to explain the sequence of basic steps for computer programs. Floyd's insight was to build upon flowcharts and to apply an invariant assertion to each branch in the flowchart. These assertions state the essential relations that exist between the variables at that point in the flow chart. It involves applying relations such as "$R = Z > 0, X = 1, Y = 0$", to each branch in the flowchart. Floyd also provided a general flowchart language to apply his method to programming languages. The language essentially contains boxes linked by flow of control arrows [Flo:67].

For example, if the assertion Q is true on entry to a branch where the condition at the branch is P. Then, the assertion on exit from the branch is $Q \wedge \neg P$ if P is false and $Q \wedge P$ otherwise (Fig. 4.6).

Another example of the use of assertions is that of the assignment statement. Suppose x represents a variable and v represents a vector consisting of all the variables in the program. Suppose $f(x, v)$ represents a function or expression of x and the other program variables represented by the vector v. Suppose the assertion $S(f(x, v), v)$ is true before the assignment $x = f(x, v)$. Then the assertion $S(x, v)$ is true after the assignment. This is represented in Floyd's flowcharts in Fig. 4.7.

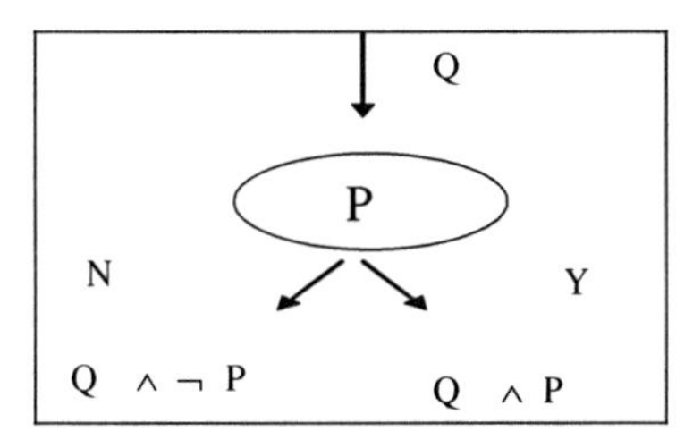

Fig. 4.6 Branch assertions in flowcharts

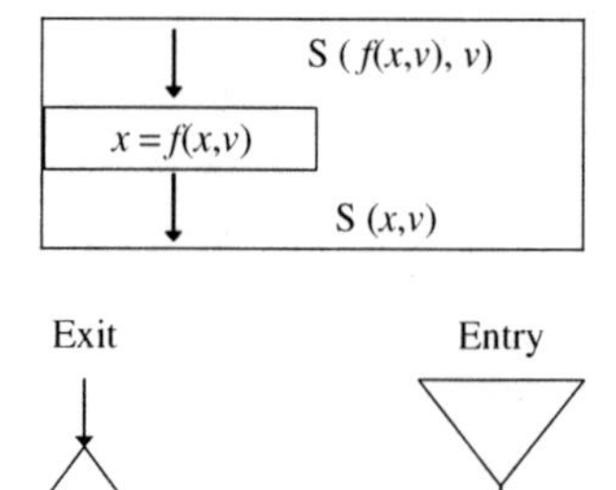

Fig. 4.7 Assignment assertions in flowcharts

Fig. 4.8 Entry and exit in flowcharts

Floyd used flowchart symbols to represent entry and exit to the flowchart. This included entry and exit assertions to describe the program's entry and exit conditions. These are defined in Fig. 4.8.

Floyd's technique shows the relationship between computer programs and a series of logical assertions. Each assertion is true whenever control passes to it. Statements in the programming language appear between the assertions. The initial assertion states the conditions that muse be true for execution of the program to take place, and the exit assertion essentially describes what must be true when the program terminates: i.e., what the program computes.

A key insight of Floyd was his recognition of the principle that if it can be shown that the assertions immediately following each step are consequences of the assertions immediately preceding it, then the assertions at the end of the program will be true, provided the appropriate assertions were true at the beginning of the program.

Floyd's simple but very powerful principle was published in "Assigning Meanings to Programs" in 1967 [Flo:67]. This paper was to influence Hoare's work on preconditions and "post-conditions, and led to what would become known as Hoare logic [Hoa:69]. Floyd's paper also presented a formal grammar for flowcharts, together with rigorous methods for verifying the effects of basic actions like assignments.

Floyd also did research work on compilers and on the translation of programming languages into machine languages. This included work on the theory of parsing [Flo:63, Flo:64], and on the theory of compilers. His work led to improved algorithms for parsing sentences and phrases in programming languages. He worked closely with Donald Knuth, and reviewed Knuth's "The Art of Computer Programming".[13] Floyd received the Turing Award in 1978 for his influence on methodologies for the creation of efficient and reliable software, and for his contribution to the theory of parsing, the semantics of programming languages, the analysis of algorithms and for program verification.

Hoare logic was developed by the well-known British computer scientist, Hoare. It is a formal system of logic used for programming semantics and for program

[13] The Art of Computer Programming [Knu:97] was originally published in three volumes. Volume 1 appeared in 1968; Volume 2 in 1969; and Volume 3 in 1973. A fourth volume is being prepared by Knuth.

verification. It was originally published in Hoare's 1969 paper "An axiomatic basis for computer programming", and has subsequently been refined by Hoare and others. Hoare's approach provides a logical methodology for precise reasoning about the correctness of computer programs. His approach uses the rigour of mathematical logic (Fig. 4.9).

Hoare's early work in computing was in the early 1960s, when he worked as a programmer at Elliott Brothers in the United Kingdom. His first assignment was the implementation of a subset of the ALGOL 60 programming language. This language had been designed by an international committee, and its concise specification was twenty-one pages long [Nau:60]. Peter Naur's report gave the implementer of the language accurate information to implement a compiler for the language, and there was no need for the implementer to communicate with the language designers.

The grammar of ALGOL 60 was specified in Backus Naur Form (BNF) and this was based on the work of Noam Chomsky. The success of BNF in specifying the syntax of ALGOL 60 led to its use in the specification of the syntax of other programming languages, and to a growth in research in the theory of formal semantics

Fig. 4.9 C.A.R. Hoare

of programming languages. One view that existed at that time was that the meaning of a language should be given by a compiler and its target implementation. Hoare argued that the meaning of a language should be independent of its implementation on the machine that the language is to run on. He preferred to avoid operational approaches to programming language semantics, and proposed instead the axiomatic approach.

Hoare became professor of Queens University in Belfast in 1968 and was influenced by Floyd's 1967 paper that applied assertions to flowcharts. Hoare recognised that this provided an effective method for proving the total correctness of programs, and he built upon Floyd's approach to cover all of the familiar constructs of high-level programming languages. This led to the axiomatic approach to defining the semantics of every statement in a programming language. The approach consists of axioms and proof rules. He introduced what has become known as the Hoare triple, and this describes how the execution of a fragment of code changes the state. A Hoare triple is of the form:

$$P \{Q\} R$$

where P and R are assertions and Q is a program or command. The predicate P is called the *precondition*, and the predicate R is called the *postcondition*.

Definition 4.2 (Partial Correctness) The meaning of the Hoare triple above is that whenever the predicate P holds of the state before the execution of the command or program Q, then the predicate R will hold after the execution of Q. The brackets indicate partial correctness as if Q does not terminate then R can be any predicate. R may be chosen to be false to express that Q does not terminate.

Total correctness requires Q to terminate, and at termination R is true. Termination needs to be proved separately. Hoare logic includes axioms and rules of inference rules for the constructs of imperative programming language. Hoare and Dijkstra were of the view that the starting point of a project should always be the specification, and that the proof of the correctness of the program should be developed along with the program itself.

4.4 Software Engineering Mathematics

The use of mathematics plays a key role in the classical engineer's work. For example, bridge designers will develop a mathematical model of a bridge prior to its construction. The model is a simplification of the reality, and an exploration of the model enables a deeper understanding of the proposed bridge to be gained. Engineers will model the various stresses on the bridge to ensure that the bridge design can deal with the projected traffic flow. The ability to use mathematics to solve practical problems is part of the engineer's education, and is part of the daily work of an engineer. The engineer applies mathematics and models to the design of the product, and the analysis of the design is a mathematical activity.

Mathematics allows a rigorous analysis to take place and avoids an over-reliance on intuition. The focus in engineering needs to be on mathematics that can be applied to solve practical problems and in developing products that are fit for use, rather than on mathematics for its own sake that is the focus of the pure mathematician. The emphasis in engineering is always in the application of the theorem rather than in the proof, and the objective is therefore to teach students how to use and apply mathematics to program well and to solve practical problems.

There is a rich body of classical mathematics available that may be applied to software engineering. Other specialized mathematical methods and notations have been developed by others to assist in software engineering (e.g., Z, VDM, VDM$^\clubsuit$, and CSP). The mathematical foundation for software engineering should include the following (Table 4.2).

The use of mathematics in software engineering is discussed in more detail in [ORg:06]. The emphasis is on mathematics that can be applied to solve practical

Table 4.2 Mathematics for software engineering

Area	Description
Set Theory	This material is elementary but fundamental.. It includes, set union and intersection operations; the Cartesian product of two sets, etc.
Relations	A relation between A and B is a subset of $A \times B$. For example, the relation T(A, A) where $(a_1, a_2) \in T$ if a_1 is taller than a_2
Functions	A function f: $A \rightarrow B$ is a relation where for each $a \in A$ there is exactly one $b \in B$ such that $(a, b) \in f$. This is denoted by $f(a) = b$. Functions may be total or partial.
Logic	Logic is the foundation for formal reasoning. It includes the study of propositional calculus and predicate calculus. It enables further truths to be derived from existing truths.
Calculus	Calculus is used extensively in engineering and physics to solve practical problems. It includes differentiation and integration, numerical methods, solving differential equations, etc.
Probability Theory	Probability theory is concerned with determining the mathematical probability of various events occurring. One example of its use in software engineering is in predicting the reliability of a software product.
Finite State Machines	Finite state machines are mathematical entities that are employed to model the execution of a program. The mathematical machine is in a given state, and depending on the input there is a change to a new state. They may be deterministic or non-deterministic.
Tabular Expressions	This approach developed by Parnas and others may be employed to specify the requirements of a system. It allows complex predicate calculus expressions to be presented in a more readable form (in a 2-dimensional table).
Graph Theory	Graphs are useful in modeling networks and a graph consists of vertices and edges. An edge joins two vertices.
Matrix Theory	This includes the study of 2×2 and $m \times n$ dimensional matrices. It includes calculating the determinants of a matrix and the inverses of a matrix.

problems rather than on theoretical mathematics. Next, we consider various formal methods that may be employed to assist in the development of high-quality software.

4.5 Formal Methods

Formal methods consist of formal specification languages or notations, and generally employ a collection of tools to support the syntax checking of the specification, as well as the proof of properties of the specification. This abstraction away from implementation enables questions about what the system does to be answered independently of the implementation, i.e., the detailed code. Furthermore, the unambiguous nature of mathematical notation avoids the problem of speculation about the meaning of phrases in an imprecisely worded natural language description of a system. Natural language is inherently ambiguous and subject to these limitations, whereas mathematics employs a precise notation with sound rules of inference. Spivey [Spi:92] defines formal specification as:

Definition 4.3 (Formal Specification) Formal specification is the use of mathematical notation to describe in a precise way the properties which an information system must have, without unduly constraining the way in which these properties are achieved.

The formal specification thus becomes the key reference point for the different parties concerned with the construction of the system. This includes determining customer needs, program implementation, testing of results, and program documentation. It follows that the formal specification is a valuable means of promoting a common understanding for all those concerned with the system. The term *formal methods* is used to describe a formal specification language and a method for the design and implementation of computer systems.

The specification is written in a mathematical language, and the implementation is derived from the specification via step-wise refinement.[14] The refinement step makes the specification more concrete and closer to the actual implementation. There is an associated proof obligation that the refinement is valid, and that the concrete state preserves the properties of the more abstract state. Thus, assuming that the original specification is correct and the proofs of correctness of each refinement step are valid, then there is a very high degree of confidence in the correctness of the implemented software. Step-wise refinement is illustrated as follows: the initial specification S is the initial model M_0, it is then refined into the more concrete model M_1, and M_1 is then refined into M_2, and so on until the eventual implementation $M_n = E$ is produced.

[14] It is questionable whether step-wise refinement is cost effective in mainstream software engineering, as it involves re-writing a specification *ad nauseum*. It is time-consuming to proceed in refinement steps with significant time also required to prove that the refinement step is valid. For safety-critical applications the main driver is quality and correctness rather than time. However, in mainstream software engineering often time to market is a key driver with a fine balance necessary between time and quality.

$$S = M_0 \subseteq M_1 \subseteq M_2 \subseteq M_3 \subseteq \ldots\ldots.. \subseteq M_n = E$$

Requirements are the foundation from which the system is built, and irrespective of the best design and development practices, the product will be incorrect if the requirements are incorrect. The objective of requirements validation is to ensure that the requirements are correct and reflect what is actually required by the customer (in order to build the right system). Formal methods may be employed to model the requirements, and the model exploration yields further desirable or undesirable properties. The ability to prove that certain properties are true of the specification is very valuable, especially in safety critical and security critical applications. These properties are logical consequences of the definition of the requirements, and, if appropriate, the requirements may need to be amended appropriately. Thus, formal methods may be employed for requirements validation and in a sense to debug the requirements.

The use of formal methods generally leads to more robust software and to increased confidence in the correctness of the software. The challenges involved in the deployment of formal methods in an organization include the education of staff in formal specification, as formal specification and the use of mathematical techniques may be a culture shock to many staff.

Formal methods have been applied to a diverse range of applications, including circuit design, artificial intelligence, the security critical field, the safety critical field, the railway sector, microprocessor verification, the specification of standards, and the specification and verification of programs.

Formal methods have been criticized by Parnas and others on the following grounds (Table 4.3).

However, formal methods are potentially quite useful in software engineering. The use of a method such as Z or VDM forces the engineer to be precise and helps to avoid ambiguities present in natural language. My experience is that formal specifications such as Z or VDM are reasonably easy to use. Clearly, a formal specification should be subject to peer review to ensure that it is correct. New formalisms may potentially add value in expressive power but need to be intuitive to be usable by practitioners. The advantage of classical mathematics is that it is familiar to students.

4.5.1 Why Should We Use Formal Methods?

There is a very strong motivation to use best practices in software engineering in order to produce software adhering to high quality standards. Flaws in software may at best cause minor irritations to customers, and in a worst-case scenario could potentially cause major damage to a customer's business or loss of life. Consequently, companies will employ best practices to mature their software processes. Formal methods are one leading-edge technology which studies suggest may be of benefit to companies who wish to minimize the occurrence of defects in software products.

Table 4.3 Criticisms of formal methods

No.	Criticism
1.	Often the formal specification (written for example in Z or VDM) is as difficult to read as the program and it therefore does not add value[15].
2.	Many formal specifications are wrong.[16]
3.	Formal methods are strong on syntax but provide little assistance in deciding on what technical information should be recorded using the syntax.[17]
4.	Formal specifications provide a model of the proposed system. However, a precise unambiguous mathematical statement of the requirements is what is needed. It should be clear what is required and what is not required.[18]
5.	Step-wise refinement is unrealistic.[19] It is like, for example, deriving a bridge from the description of a river and the expected traffic on the bridge. Refining a formal specification to a program is like refining a blueprint until it turned into a house. This is hardly realistic and there is always a need for a creative step in design.
6.	Many unnecessary mathematical formalisms have been developed in formal methods rather than using the available classical mathematics.[20]

The use of formal methods is mandatory in certain circumstances. The Ministry of Defence in the United Kingdom issued two safety-critical standards in the early 1990s related to the use of formal methods in the software development lifecycle. The first is Defence Standard 0055, i.e., Def Stan 00-55, "The Procurement of safety critical software in defense equipment" [MOD:91a]. This standard makes it mandatory to employ formal methods in safety-critical software development in the United Kingdom; and mandates the use of formal proof that the most crucial programs correctly implement their specifications. The other Defence Standard is Def Stan 00-56 "Hazard analysis and safety classification of the computer and programmable

[15] Of course, others might reply by saying that some of Parnas's tables are not exactly intuitive, and that the notation he employs in some of his tables is quite unfriendly. The usability of all of the mathematical approaches needs to be enhanced if they are to be taken seriously by industrialists.

[16] Obviously, the formal specification must be analyzed using mathematical reasoning and tools to ensure its correctness. The validation of a formal specification can be carried out using mathematical proof of key properties of the specification; software inspections; or specification animation.

[17] Approaches such as VDM include a method for software development as well as the specification language.

[18] Models are extremely valuable as they allow simplification of the reality. A mathematical study of the model demonstrates whether it is a suitable representation of the system. Models allow properties of the proposed requirements to be studied prior to implementation.

[19] Step-wise refinement involves rewriting a program *ad nauseum* with each refinement step produces a more concrete specification (that includes code and formal specification) until eventually the detailed code is produced. However, tool support may make refinement easier. The refinement calculus offers a very rigorous approach to develop correct programs from formal specifications. However, it is debatable whether it is cost effective in mainstream software development.

[20] My preference is for the use of classical mathematics for specification. However, I see approaches such as VDM or Z as useful as they do add greater rigour to the software development process than is achieved with the use of natural language. In fact, both Z and VDM are reasonably easy to learn, and there have been some good results obtained by their use. I prefer to avoid the fundamentalism that I have seen exhibited elsewhere, and am of the view that if other notations add value in formal specification then it is perfectly reasonable to employ them.

electronic system elements of defense equipment" [MOD:91b]. The objective of this standard is to provide guidance to identify which systems or parts of systems being developed are safety-critical and thereby require the use of formal methods. This proposed system is subject to an initial hazard analysis to determine whether there are safety-critical parts. The reaction to these defence standards 00-55 and 00-56[21] was quite hostile initially, as most suppliers were unlikely to meet the technical and organization requirements of the standard. This is described in [Tie:91] and Brown in [Bro:90] argues that

Comment 4.1 (Missile Safety) Missile systems must be presumed dangerous until shown to be safe, and that the absence of evidence for the existence of dangerous errors does not amount to evidence for the absence of danger.

It is quite possible that a software company may be sued for software which injures a third party, and it is conjectured in [Mac:93] that the day is not far off when

> A system failure traced to a software fault, and injurious to a third party, will lead to a successful litigation against the developers of the said system software.

This suggests that companies will need a quality assurance program that will demonstrate that every reasonable practice was considered to prevent the occurrence of defects. One such practice for defect prevention is the use of formal methods in the software development lifecycle, and in some domains, e.g., the safety critical domain, it looks likely that the exclusion of formal methods in the software development cycle may need to be justified.

There is evidence to suggest that the use of formal methods provides savings in the cost of the project, for example, an independent audit of the large CICS transaction processing project at IBM demonstrated a 9% cost saving attributed to the use of formal methods. An independent audit of the Inmos floating point unit of the T800 project confirmed that the use of formal methods led to an estimated 12-month reduction in testing time. These savings are discussed in more detail in chapter one of [HB:95].

The approach of modern software companies to providing high-quality software on time and within budget is to employ a mature software development process including inspections and testing. Models such as the CMM [Pau:93], its successor, the CMMI [CKS:03], and ISO 9000 are employed to assist the organization to mature its software process. The process-based approach is also useful in that it demonstrates that reasonable practices are employed to identify and prevent the occurrence of defects, and an ISO 9000 or CMMI certified software development organization has been independently assessed to have reasonable software development practices in place. A formal methods approach is complementary to these models, and it fits comfortably into the defect prevention key process area and the technology change management key process area on the Capability Maturity Model.

[21] The United Kingdom Defence Standards 0055 and 0056 have been revised in recent years to be less prescriptive on the use of formal methods.

4.5.2 Applications of Formal Methods

Formal methods are used in academia and to varying degrees in industry. The safety-critical and security critical fields are two key areas to which formal methods have been successfully applied. Several organizations have piloted formal methods with varying degrees of success. These include IBM, who developed VDM at its laboratory in Vienna. Another IBM site, IBM (Hursley) piloted the Z formal specification language in the United Kingdom for the CICS (Customer Information Control System) project. This is an online transaction processing system with over 500,000 lines of code. The project generated valuable feedback to the formal methods community, and although it was very successful in the sense that an independent audit verified that the use of formal methods generated a 9% cost saving, there was resistance to the deployment of the formal methods in the organization.[22] This was attributed to the lack of education on formal methods in computer science curricula, lack of adequate support tools for formal methods, and the difficulty that the programmers had with mathematics and logic.

The mathematical techniques developed by Parnas (i.e., requirements model and tabular expressions) have been employed to specify the requirements of the A-7 aircraft as part of a software cost reduction program for the United States Navy.[23] Tabular expressions have also been employed for the software inspection of the automated shutdown software of the Darlington Nuclear power plant in Canada.[24] These are two successful uses of mathematical techniques in software engineering.

Formal methods have been successfully applied to the hardware verification field; for example, parts of the Viper microprocessor[25] were formally verified, and the FM9001 microprocessor was formally verified by the Boyer Moore theorem prover [HB:95]. There are many examples of the use of formal methods in the railway domain, and examples dealing with the modeling and verification of a railroad gate

---

[22] I recall a keynote presentation by Peter Lupton of IBM (Hursley) at the Formal Methods Europe (FME'93) conference in which he noted that there was a resistance to the use of formal methods among the software engineers at IBM (Hursley), and that the engineers found the Z notation to be slightly unintuitive.

[23] However, I was disappointed to learn that the resulting software was actually never deployed on the A-7 aircraft, and that it was essentially just a research project. Similarly, the time and cost involved in the mathematical verification of the Darlington nuclear power plant was a concern to managerment.

[24] This was an impressive use of mathematical techniques and it has been acknowledged that formal methods must play an important role in future developments at Darlington. However, given the time and cost involved in the software inspection of the shutdown software some managers have less enthusiasm in shifting from hardware to software controllers [Ger:94].

[25] The VIPER microprocessor chip was very controversial. It is an example of where formal methods were oversold in that the developers RSRE (Royal Signals and Radar Establishment) of the United Kingdom and Charter (a company licensed to exploit the VIPER results) claimed that the VIPER chip was formally specified and that the chip design conforms to the specification. However, the report by Avra Cohen of Cambridge University's computer laboratories argued that the claim by RSRE and Charter was misleading. Computational Logic of the United States later confirmed Avra Cohn's conclusions.

controller and railway signaling are described in [HB:95]. Clearly, it is essential to verify safety critical properties such as "when the train goes through the level crossing then the gate is closed". The mandatory use of formal methods in some safety and security-critical fields has led to formal methods being employed to verify correctness in the nuclear power industry, in the aerospace industry, in the security technology area, and the railroad domain. These sectors are subject to stringent regulatory controls to ensure safety and security.

Formal methods have been successfully applied to the telecommunications domain, and have been useful in investigating the feature interaction problem as described in [Bou:94]. The EC SCORE project considered mathematical techniques to identify and eliminate feature interaction in the telecommunications environment. The feature interaction problem occurs when two features that work correctly in isolation fail to work correctly together.

Formal methods have been applied to domains which have little to do with computer science, for example, to the problem of the formal specification of the single transferable voting system in [Pop:97], and to various organizations and structures in [ORg:97]. There is a collection of examples to which formal methods have been applied in [HB:95].

4.5.3 Tools for Formal Methods

One key criticism of formal methods is the lack of available or usable tools to support the engineer in writing the formal specification or in doing the proof. Many of the early tools were criticized as being of academic use only, and not being of industrial strength. However, in recent years better tools have become available, and it is likely that more advanced tools will become available in the coming years to support the engineer's work in formal specification and formal proof.

There are various kinds of tools employed to support the formal software development environment. These include syntax checkers to check that the specification is syntactically correct; specialized editors to ensure that the written specification is syntactically correct; tools to support refinement; automated code generators to generate a high-level language corresponding to the specification; theorem provers to demonstrate the presence or absence of key properties and to prove the correctness of refinement steps, and to identify and resolve proof obligations; and specification animation tools where the execution of the specification can be simulated. Such tools are available from vendors like B-Core and IFAD.

The tools are developed to support existing methods and the trend is towards an integrated set of method and tools rather than loosely coupled tools. For example, the B-Toolkit from B-Core is an integrated set of tools that supports the B-Method. These include syntax and type checking, specification animation, proof obligation generator, an auto-prover, a proof assistor, and code generation. Thus, in theory, a complete formal development from initial specification to final implementation may be achieved, with every proof obligation justified, leading to a provably correct program.

The IFAD Toolbox[26] is a well-known support tool for the VDM-SL specification language, and it includes support for syntax and type checking, an interpreter and debugger to execute and debug the specification, and a code generator to convert from VDM-SL to C++. It also includes support for graphical notations such as the OMT/UML design notations.

SDL is a specification language which is employed in event driven real time systems. It is an object-orientated graphical formal language, and support for SDL is provided by the SDT tool from Telelogic. The SDT tool provides code generation from the specification into the C or C++ programming languages, and the generated code can be used in simulations as well as in applications. Telelogic provides an ITEX tool which may be used with or independently of SDT. It allows the generation of a test suite from the SDL specification, thereby speeding up the testing cycle.

The RAISE tools are an integrated toolset including the RAISE specification language (RSL) and a collection of tools to support software development including editors and translators from the specification language to Ada or C++. There are many other tools available, including the Boyer Moore theorem prover, the FDR tool for CSP, the CADiZ tool for the Z specification language, the Mural tool for VDM, the LOTOS toolbox for LOTOS specifications, and the PVS tool.

Finally, various research groups are investigating methods and tools to assist the delivery of high-quality software. This includes a group led by Parnas at SQRL at the University of Limerick in Ireland,[27] and this group is developing a table tool system to support tabular expressions which they believe will support engineers in specifying requirements. The suite of tools include tools for the creation of tables; tools to check consistency and completeness properties of tables; tools to perform table composition and tools to generate a test oracle from a table.

4.5.3.1 Formal Methods and Reuse

Effective software reuse helps to improve software development productivity, and this is important in rapid application software development. The idea is to develop building blocks which may then be reused in future projects, and this requires high quality and reliable components. It is essential to identify the domains to which the component may be applied, and that a documented description exists of the actual behavior of the component and the circumstances in which it may be employed.[28]

---

[26] The IFAD Toolbox has been renamed to VDMTools as IFAD sold the VDM Tools to CSK in Japan. The tools are expected to be available worldwide and will be improved further.

[27] This group is being supported by Science Foundation Ireland with €5–6 million of Irish taxpayers' funds. It remains to be seen whether the results will be of benefit to industrialists.

[28] There is a lot of reusable software that has been developed but never reused. This is a key challenge that companies have to face if they wish to reduce their development costs, and deliver their products to market faster. Reducing development costs and faster delivery of products are two key drivers in today's competitive environment.

Effective reuse is typically limited to a particular domain, and there are reuse models to assist to assess the current reuse practices in the organization. This enables a reuse strategy to be developed and implemented. Systematic reuse is being researched in academia and industry, and the ROADS project was an EC funded project which included the European Software Institute (ESI) and Thompson as partners to investigate a reuse approach for domain-based software. The software product line approach [ClN:02] is proposed by the SEI. Formal methods have a role to play in software reuse also, as they offer enhanced confidence in the correctness of the component, as well as providing an unambiguous formal description of the behavior of the particular component. The component may be tested extensively to provide extra confidence in its correctness. A component is generally used in different environments, and the fact that a component has worked successfully in one situation is no guarantee that it will work successfully in the future, as there could be potential undesirable interaction between it and other components, or other software. Consequently, it is desirable that the behavior of the component be unambiguously specified and fully understood, and that a formal analysis of component composition be performed to ensure that risks are minimized, and that the resulting software is of a high quality.

There has been research into the formalization of components in both academia and industry. The EC funded SCORE research project conducted as part of the European RACE II program considered the challenge of reuse. It included the formal specification of components and developed a component model. Formal methods have a role to play in identifying and eliminating undesirable component interaction.

4.5.4 Model-Oriented Approach

There are two key approaches to formal methods: namely the model-oriented approach of VDM or Z, and the algebraic, or axiomatic approach, which includes the process calculi such as the calculus communicating systems (CCS) or communicating sequential processes (CSP).

A model-oriented approach to specification is based on mathematical models. A mathematical model is a mathematical representation or abstraction of a physical entity or system. The representation or model aims to provide a mathematical explanation of the behavior of the system or the physical world. A model is considered suitable if its properties closely match the properties of the system, and if its calculations match and simplify calculations in the real system, and if predictions of future behavior may be made. The physical world is dominated by models, e.g., models of the weather system, that enable predictions of the weather to be made, and economic models that enable predictions of the future performance of the economy may be made.

It is fundamental to explore the model and to consider the behavior of the model and the behavior of the physical world entity. The adequacy of the model is the extent to which it explains the underlying physical behavior, and allows predictions of future behavior to be made. This will determine its acceptability as a representation of the physical world. Models that are ineffective will be replaced with newer models which offer a better explanation of the manifested physical behavior. There are many examples in science of the replacement of one theory by a newer one. For example, the Copernican model of the universe replaced the older Ptolemaic model, and Newtonian physics was replaced by Einstein's theories on relativity. The structure of the revolutions that take place in science are described in [Kuh:70].

A model is a foundation stone from which the theory is built, and from which explanations and justification of behavior are made. It is not envisaged that we should justify the model itself, and if the model explains the known behavior of the system, it is thus deemed adequate and suitable. Thus the model may be viewed as the starting point of the system. Conversely, if inadequacies are identified with the model we may view the theory and its foundations as collapsing, in a similar manner to a house of cards; alternately, amendments to the theory to address the inadequacies may be sought.

The model-oriented approach to software development involves defining an abstract model of the proposed software system. The model acts as a representation of the proposed system, and the model is then explored to assess its suitability. The exploration of the model takes the form of model interrogation, i.e., asking questions and determining the effectiveness of the model in answering the questions. The modeling in formal methods is typically performed via elementary discrete mathematics, including set theory, sequences, functions and relations.

The modeling approach is adopted by the Vienna Development Method (VDM) and Z. VDM arose from work done in the IBM laboratory in Vienna in formalizing the semantics for the PL/1 compiler, and it was later applied to the specification of software systems. The Z specification language had its origins in work done at Oxford University in the early 1980s.

VDM includes a methodology for software development as well as its specification language originally named Meta IV (a pun on *metaphor*), and later renamed to VDM-SL in the standardization of VDM. The approach to software development is via step-wise refinement. There are several schools of VDM, including VDM^{++}, the object-oriented extension to VDM, and what has become known as the Irish school of VDM, i.e., $\text{VDM}^{\clubsuit}$, which was developed at Trinity College, Dublin.

4.5.5 Axiomatic Approach

The axiomatic approach focuses on the properties that the proposed system is to satisfy, and there is no intention to produce an abstract model of the system. The required properties and behavior of the system are stated in mathematical notation. The difference between the axiomatic specification and a model-based approach is illustrated by the example of a stack. The stack includes operators for pushing an element onto the stack and popping an element from the stack. The properties of *pop* and *push* are explicitly defined in the axiomatic approach. The model-oriented approach constructs an explicit model of the stack and the operations are defined in terms of the effect that they have on the model. The specification of an abstract data type of a stack involves the specification of the properties of the abstract data type. However, the abstract data type is not explicitly defined; i.e., only the properties are defined. The specification of the *pop* operation on a stack is given by axiomatic properties, for example, $pop(push(s, \ x)) = s$.

Comment 4.2 (Axiomatic Approach) The property-oriented approach has the advantage that the implementer is not constrained to a particular choice of implementation, and the only constraint is that the implementation must satisfy the stipulated properties.

The emphasis is on the identification and expression of the required properties of the system, and the actual representation or implementation issues are avoided. That is, the focus is on the specification of the underlying behavior. Properties are typically stated using mathematical logic or higher-order logics, and mechanized theorem-proving techniques may be employed to prove results.

One potential problem with the axiomatic approach is that the properties specified may not be satisfiable in any implementation. Thus, whenever a "formal axiomatic theory" is developed a corresponding "model" of the theory must be identified, in order to ensure that the properties may be realized in practice. That is, when proposing a system that is to satisfy some set of properties, there is a need to prove that there is at least one system that will satisfy the set of properties. The model-oriented approach has an explicit model to start with and so this problem does not arise.

A constructive approach is preferred by some groups in formal methods, and in this approach whenever existence is stipulated *constructive existence* is implied, where a direct example of the existence of an object can be exhibited, or an algorithm to produce the object within a finite time period exists. This is different from an existence proof, where it is known that there is a solution to a particular problem, but where there is no algorithm to construct the solution.

4.5.6 The Vienna Development Method

VDM dates from work done by the IBM research laboratory in Vienna. The aim of this group was to specify the semantics of the PL/1 programming language. This was achieved by employing the Vienna Definition Language (VDL) and the group took an operational semantic approach. That is, the semantics of the language is determined in terms of a hypothetical machine which interprets the programs of that language [BjJ:78, BjJ:82]. Later work led to the Vienna Development Method (VDM) with its specification language, Meta IV. This concerned itself with the denotational semantics of programming languages; i.e., a mathematical object (set, function, etc.) is associated with each phrase of the language [BjJ:82]. The mathematical object is termed the *denotation* of the phrase.

VDM is a *model-oriented approach* and this means that an explicit model of the state of an abstract machine is given, and operations are defined in terms of this state. Operations may act on the system state, taking inputs, and producing outputs as well as a new system state. Operations are defined in a precondition and post-condition style. Each operation has an associated proof obligation to ensure that if the precondition is true, then the operation preserves the system invariant. The initial state itself is, of course, required to satisfy the system invariant. VDM uses keywords to distinguish different parts of the specification, e.g., preconditions, post-conditions, as introduced by the keywords *pre* and *post* respectively. In keeping with the philosophy that formal methods specifies *what* a system does as distinct from *how*, VDM employs post-conditions to stipulate the effect of the operation on the state. The previous state is then distinguished by employing *hooked variables*, e.g., v¬, and the postcondition specifies the new state (defined by a logical predicate relating the pre-state to the post-state) from the previous state.

VDM is more than its specification language Meta IV (called VDM-SL in the standardization of VDM), and it is, in fact, a software development method, with rules to verify the steps of development. The rules enable the executable specification, i.e., the detailed code, to be obtained from the initial specification via refinement steps. Thus, we have a sequence $S = S_0, S_1, \ldots, S_n = E$ of specifications, where S is the initial specification, and E is the final (executable) specification. Retrieval functions enable a return from a more concrete specification to the more abstract specification. The initial specification consists of an initial state, a system state, and a set of operations. The system state is a particular domain, where a domain is built out of primitive domains such as the set of natural numbers, etc., or constructed from primitive domains using domain constructors such as Cartesian product, disjoint union, etc. A domain-invariant predicate may further constrain the domain, and a *type* in VDM reflects a domain obtained in this way. Thus, a type in VDM is more specific than the signature of the type, and thus represents values in the domain defined by the signature, which satisfy the domain invariant. In view of this approach to types, it is clear that VDM types may not be "statically type checked".

VDM specifications are structured into modules, with a module containing the module name, parameters, types, operations, etc. Partial functions occur frequently in computer science as many functions, especially recursively defined functions can be undefined, or fail to terminate for some arguments in their domain. VDM addresses partial functions by employing nonstandard logical operators, namely the logic of partial functions (LPFs) which can deal with undefined operands. The Boolean expression $T \vee \bot = \bot \vee T = T$; i.e., the truth value of a logical or operation is true if at least one of the logical operands is true, and the undefined term is treated as a don't care value.

VDM has been used in industrial strength projects, as well as by the academic community. There is tool support available, for example, the IFAD Toolbox.[29] VDM is described in more detail in [ORg:06]. There are several variants of VDM,

[29] As discussed earlier the VDM Tools from the IFAD Toolbox are now owned by the CSK Group in Japan.

including VDM^{++}, the object-oriented extension of VDM, and the Irish school of the VDM, which is discussed in the next section.

4.5.7 VDM♣, the Irish School of VDM

The Irish School of VDM is a variant of standard VDM, and is characterized by [Mac:90] its constructive approach, classical mathematical style, and its terse notation. In particular, this method aims to combine the *what* and *how* of formal methods in that its terse specification style stipulates in concise form *what* the system should do; furthermore, the fact that its specifications are constructive (or functional) means that the *how* is included with the *what*. However, it is important to qualify this by stating that the how as presented by VDM♣ is not directly executable, as several of its mathematical data types have no corresponding structure in high-level programming languages or functional languages. Thus, a conversion or reification of the specification into a functional or higher-level language must take place to ensure a successful execution. It should be noted that the fact that a specification is constructive is no guarantee that it is a good implementation strategy, if the construction itself is naive. This issue is considered (cf. pp. 135–7 in [Mac:99]) in the construction of the Fibonacci series.

The Irish school follows a similar development methodology as in standard VDM, and is a model-oriented approach. The initial specification is presented, with initial state and operations defined. The operations are presented with preconditions; however, no postcondition is necessary as the operation is "functionally" (i.e., explicitly) constructed. Each operation has an associated proof obligation: i.e., if the precondition for the operation is true, and the operation is performed, then the system invariant remains true after the operation. The proof of invariant preservation normally takes the form of *constructive proofs*. This is especially the case for *existence proofs*, in that the philosophy of the school is to go further than to provide a theoretical proof of existence; rather the aim is to exhibit existence constructively.

The emphasis is on constructive existence, and the implication of this is that the school avoids the existential quantifier of predicate calculus. In fact, reliance on logic in proof is kept to a minimum, and emphasis instead is placed on equational reasoning. Special emphasis is placed on studying algebraic structures and their morphisms. Structures with nice algebraic properties are sought, and such a structure includes the monoid, which has closure, associativity, and a unit element. The monoid is a very common structure in computer science. The concept of isomorphism is powerful, reflecting that two structures are essentially identical, and thus we may choose to work with either, depending on which is more convenient for the task in hand.

The school has been influenced by the work of Polya and Lakatos. The former [Pol:57] advocated a style of problem solving characterized by first considering an easier sub-problem, and considering several examples. This generally leads to a clearer insight into solving the main problem. Lakatos's approach to mathematical discovery [Lak:76] is characterized by heuristic methods. A primitive conjecture is proposed and if global counter-examples to the statement of the conjecture are discovered, then the corresponding *hidden lemma* for which this global counterexample is a local counter example is identified and added to the statement of the primitive conjecture. The process repeats, until no more global counterexamples are found. A skeptical view of absolute truth or certainty is inherent in this.

Partial functions are the norm in VDM$^\clubsuit$, and as in standard VDM, the problem is that recursively defined functions may be undefined, or fail to terminate for several of the arguments in their domain. The logic of partial functions (LPFs) is avoided, and instead care is taken with recursive definitions to ensure termination is achieved for each argument. This is achieved by ensuring that the recursive argument is strictly decreasing in each recursive invocation. The $\perp$ symbol is typically used in the Irish school to represent *undefined or unavailable* or *do not care*. Academic and industrial projects have been conducted using the method of the Irish school, but at this stage tool support is limited.

There are proof obligations to demonstrate that the operations preserve the invariant. Proof obligations require a mathematical proof by hand or a machine-assisted proof to verify that the invariant remains satisfied after the operation.

4.5.8 The Z Specification Language

Z is a formal specification language founded on Zermelo set theory. It was developed in the late 1970s and early 1980s by Jean-Raymond Abrial at the Programming Research Group at Oxford. It is used for the formal specification of software and is a model-oriented approach. An explicit model of the state of an abstract machine is given, and the operations are defined in terms of the effect that they have on the state. The main features of the language include a mathematical notation which is similar to VDM and the schema calculus. The latter is visually striking and consists essentially of boxes, with these boxes or schemas used to describe operations and states. The schema calculus enables schemas to be used as building blocks and combined with other schemas. The *Z* specification language was published as an ISO standard (ISO/IEC 13568:2002) in 2002.

The schema calculus is a powerful means of decomposing a specification into smaller pieces or schemas. This decomposition helps to ensure that a *Z* specification is highly readable, as each individual schema is small in size and self-contained. Exception handling may be addressed by defining schemas for the exception cases, and then combining the exception schema with the original operation schema. Mathematical data types are used to model the data in a system and these data types obey mathematical laws. These laws enable simplification of expressions and are useful with proofs.

Operations are defined in a precondition/postcondition style. However, the precondition is implicitly defined within the operation; i.e., it is not separated out as in standard VDM. Each operation has an associated proof obligation to ensure that if the precondition is true, then the operation preserves the system invariant. The initial state itself is, of course, required to satisfy the system invariant. Postconditions employ a logical predicate which relates the pre-state to the post-state, the post-state of a variable being distinguished by priming, e.g., v'. Various conventions are employed within Z specification, for example $v?$ indicates that v is an input variable; v' indicates that v is an output variable. The symbol Ξ *Op* operation indicates that the operation *Op* does not affect the state, whereas the symbol Δ *Op* indicates that *Op* is an operation which affects the state.

Many of the data types employed in Z have no counterpart in standard programming languages. It is therefore important to identify and describe the concrete data structures which ultimately will represent the abstract mathematical structures. As the concrete structures may differ from the abstract, the operations on the abstract data structures may need to be refined to yield operations on the concrete data which yield equivalent results. For simple systems, direct refinement (i.e., one step from abstract specification to implementation) may be possible; in more complex systems, deferred refinement is employed, where a sequence of increasingly concrete specifications are produced to yield the executable specification eventually.

Z has been successfully applied in industry, and one of its well-known successes is the CICS project at IBM Hursley in England. A more detailed account of Z is in [ORg:06, Wrd:92]

4.5.8.1 The B-Method

The *B-Technologies* [McD:94] consist of three components: a method for software development, namely the *B*-Method; a supporting set of tools, namely, the *B*-Toolkit; and a generic program for symbol manipulation, namely, the *B*-Tool (from which the *B*-Toolkit is derived). The *B*-Method is a model-oriented approach and is closely related to the Z specification language. The specification language was developed by Jean-Raymond Abrial. Every construct in the method has a set theoretic counterpart, and the method is founded on Zermelo set theory. Each operation has an explicit precondition, and an immediate proof obligation is that the precondition is stronger than the weakest precondition for the operation.

One key purpose [McD:94] of the *abstract machine* in the *B*-Method is to provide encapsulation of variables representing the state of the machine and operations which manipulate the state. Machines may refer to other machines, and a machine may be introduced as a refinement of another machine. The abstract machines are specification machines, refinement machines, or implementable machines. The *B*-Method adopts a layered approach to design where the design is gradually made more concrete by a sequence of design layers. Each design layer is a refinement that involves a more detailed implementation in terms of abstract machines of the previous layer. The design refinement ends when the final layer is implemented purely in terms of library machines. Any refinement of a machine by another has associated proof obligations, and proof is required to verify the validity of the refinement step.

Specification animation of the Abstract Machine Notation (AMN) specification is possible with the *B*-Toolkit, and this enables typical usage scenarios of the AMN specification to be explored for requirements validation. This is, in effect, an early form of testing, and it may be used to demonstrate the presence or absence of desirable or undesirable behavior. Verification takes the form of a proof to demonstrate that the invariant is preserved when the operation is executed within its precondition, and this is performed on the AMN specification with the B-Toolkit.

The B-Toolkit provides several tools which support the B-Method, and these include syntax and type checking; specification animation, proof obligation generator, auto prover, proof assistor, and code generation. Thus, in theory, a complete formal development from initial specification to final implementation may be achieved, with every proof obligation justified, leading to a provably correct program.

The B-Method and toolkit have been successfully applied in industrial applications, and one of the projects to which they have been applied is the CICS project at IBM Hursley in the United Kingdom. The B-Method and Toolkit have been designed to support the complete software development process from specification to code. The application of B to the CICS project is described in [Hoa:95], and the automated support provided has been cited as a major benefit of the application of the B-Method and the B-Toolkit.

4.5.9 Propositional and Predicate Calculus

Propositional calculus associates a truth-value with each proposition and is widely employed in mathematics and logic. There are a rich set of connectives employed in the calculus for truth functional operations, and these include $A \Rightarrow B$, $A \wedge B$, $A \vee B$ which denote, respectively, the conditional of A and B, the conjunction of A and B, and the disjunction of A and B. A truth table may be constructed to show the results of these operations on the binary values of A and B. That is, A and B have the binary truth values of *true* and *false*, and the result of the truth functional operation is to yield a binary value. There are other logics that allow more than two truth values. These include, for example, the logic of partial functions which is a 3-valued logic. This logic allows a third truth value (the undefined truth-value) for the proposition as well as the standard binary values of true and false.

Predicate calculus includes variables, and a formula in predicate calculus is built up from the basic symbols of the language. These symbols include variables; predicate symbols, including equality; function symbols, including the constants; logical symbols, e.g., $\exists$, $\wedge$, $\vee$, $\neg$, etc.; and the punctuation symbols, e.g., brackets and commas. The formulae of predicate calculus are built from terms, where a *term* is a key construct, and is defined recursively as a variable or individual constant or as some function containing terms as arguments. A formula may be an atomic formula or built from other formulae via the logical symbols. Other logical symbols are then defined as abbreviations of the basic logical symbols.

An interpretation gives meaning to a formula. If the formula is a sentence (i.e., it does not contain any free variables), then the given interpretation is true or false. If a formula has free variables, then the truth or falsity of the formula depends on the values given to the free variables. A formula with free variables essentially describes a relation say, $R(x_1, x_2, \ldots, x_n)$ such that $R(x_1, x_2, \ldots, x_n)$ is true if

$(x_1, x_2, \ldots, x_n)$ is in relation R. If a formula with free variables is true irrespective of the values given to the free variables, then the formula is true in the interpretation.

A valuation function is associated with the interpretation, and this gives meaning to the formulae in the language. Thus, associated with each constant c is a constant c_Σ in some universe of values Σ; with each function symbol f, we have a function symbol f_Σ in Σ; and for each predicate symbol P a relation P_Σ in Σ. The valuation function, in effect, gives a semantics to the language of the predicate calculus L. The truth of a proposition P is then defined in the natural way, in terms of the meanings of the terms, the meanings of the functions, predicate symbols, and the normal meanings of the connectives.

Mendelson [Men:87] provides a rigorous though technical definition of truth in terms of satisfaction (with respect to an interpretation M). Intuitively, a formula F is *satisfiable* if it is *true* (in the intuitive sense) for some assignment of the free variables in the formula F. If a formula F is satisfied for every possible assignment to the free variables in F, then it is *true* (in the technical sense) for the interpretation M. An analogous definition is provided for *false* in the interpretation M.

A formula is *valid* if it is true in every interpretation; however, as there may be an uncountable number of interpretations, it may not be possible to check this requirement in practice. M is said to be a model for a set of formulae if and only if every formula is true in M.

There is a distinction between proof theoretic and model theoretic approaches in predicate calculus. *Proof theoretic* is essentially syntactic, and we have a list of axioms with rules of inference. In this way the theorems of the calculus may be logically derived, and thus we may logically derive (i.e., |-A) the theorems of the calculus. In essence the logical truths are a result of the syntax or form of the formulae, rather than the *meaning* of the formulae. *Model theoretical*, in contrast is essentially semantic. The truths derive essentially from the meaning of the symbols and connectives, rather than the logical structure of the formulae. This is written as $|-_M A$.

A calculus is *sound* if all the logically valid theorems are true in the interpretation, i.e., proof theoretic $\Rightarrow$ model theoretic. A calculus is complete if all the truths in an interpretation are provable in the calculus, i.e., model theoretic $\Rightarrow$ proof theoretic. A calculus is *consistent* if there is no formula A such that |- A and |- $\neg$ A.

4.5.9.1 Predicate Transformers and Weakest Preconditions

The precondition of a program S is a predicate, i.e., a statement that may be true or false, and it is usually required to prove that if Q is true, where Q is the precondition of a program S; i.e., $(\{Q\}S\{R\})$, then execution of S is guaranteed to terminate in a finite amount of time in a state satisfying R.

The weakest precondition (cf. p. 109 of [Gri:81]) of a command S with respect to a postcondition R represents the set of all states such that if execution begins in any one of these states, then execution will terminate in a finite amount of time in a state with R true. These set of states may be represented by a predicate Q', so that $wp(S, R) = wp_S(R) = Q'$, and so wp_S is a predicate transformer, i.e., it may be regarded as a function on predicates. The weakest precondition is the precondition that places the fewest constraints on the state than all of the other preconditions of (S, R). That is, all of the other preconditions are stronger than the weakest precondition.

The notation $Q\{S\}R$ is used to denote partial correctness and indicates that if execution of S commences in any state satisfying Q, and if execution terminates, then the final state will satisfy R. Often, a predicate Q which is stronger than the weakest precondition $wp(S, R)$ is employed, especially where the calculation of the weakest precondition is nontrivial. Thus, a stronger predicate Q such that $Q \Rightarrow wp(S, R)$ is sometimes employed in these cases.

There are many properties associated with the weakest preconditions, and these are used in practice to simplify expressions involving weakest preconditions, and in determining the weakest preconditions of various program commands, e.g., assignments, iterations, etc. These are discussed in more detail in [ORg:06]. Weakest preconditions are useful in developing a proof of correctness of a program in parallel with its development.

An imperative program may be regarded as a predicate transformer. This is since a predicate P characterises the set of states in which the predicate P is true, and an imperative program may be regarded as a binary relation on states, which may be extended to a function F, leading to the Hoare triple $P\{F\}Q$. That is, the program F acts as a predicate transformer. The predicate P may be regarded as an input assertion, i.e., a Boolean expression which must be true before the program F is executed. The Boolean expression Q is the output assertion, and is true if the program F terminates, having commenced in a state satisfying P.

4.5.9.2 The Process Calculi

The objectives of the process calculi [Hor:85] are to provide mathematical models which provide insight into the diverse issues involved in the specification, design, and implementation of computer systems which continuously act and interact with their environment. These systems may be decomposed into sub-systems which interact with each other and their environment. The basic building block is the *process*, which is a mathematical abstraction of the interactions between a system and its environment. A process which lasts indefinitely may be specified recursively. Processes may be assembled into systems, execute concurrently, or communicate with each other. Process communication may be synchronized, and this generally takes the form of a process outputting a message simultaneously to another process inputting a message. Resources may be shared among several processes. Process calculi enrich the understanding of communication and concurrency, and elegant formalisms such as CSP [Hor:85] and CCS [Mil:89] have been developed. These calculi obey a rich collection of mathematical laws.

The expression $(a?P)$ in CSP describes a process which first engages in event a, and then behaves as process P. A recursive definition is written as $(\mu X) \bullet F(X)$ and an example of a simple chocolate vending machine is:

$$\text{VMS} = \mu X : \{\text{coin, choc}\} \bullet (\text{coin?(choc?X)})$$

The simple vending machine has an alphabet of two symbols, namely, *coin* and *choc*. The behaviour of the machine is that a coin is entered into the machine, and then a chocolate selected and provided.

CSP processes use channels to communicate values with their environment, and input on channel c is denoted by $(c?.x\, P_x)$. This describes a process that accepts any value x on channel c, and then behaves as process P_x. In contrast, $(c!e\,P)$ defines a process which outputs the expression e on channel c and then behaves as process P.

The π-calculus is based on names. Communication between processes takes place between known channels, and the name of a channel may be passed over a channel. There is no distinction between channel names and data values in the π-calculus, and this is a difference between it and CCS. The output of a value v on channel a is given by $\bar{a}v$; i.e., output is a negative prefix. Input on a channel a is given by $a(x)$, and is a positive prefix. Private links or restrictions are given by $(x)P$ in the π-calculus and $P \backslash x$ in CCS.

4.5.10 Finite State Machines

Early work on finite state automata was published by Warren McCulloch and Walter Pitts, in 1943. They were neurophysiologists and their paper "A Logical Calculus Immanent in Nervous Activity" [McP:43] contributed to the study of neural networks as well as to the theory of automata and cybernetics. They were interested in modelling the thought process for humans and machines.

George Mealy and Edward Moore generalised the theory to more powerful machines in separate papers in the mid 1950s. These finite-state machines are referred to as the "Mealy machine" and the "Moore machine". The Mealy machine determines its outputs through the current state and the input, whereas the output of Moore's machine is based upon the current state alone.

Definition 4.4 (Finite State Machine) A finite state machine (FSM) is an abstract mathematical machine that consists of a finite number of states. It includes a start state q_0 in which the machine is in initially, a finite set of states Q, an input alphabet Σ, a state transition function δ, and a set of final accepting states F (where $F \subseteq, Q$).

The state transition function takes the current state and an input and returns the next state. That is, the transition function is of the form:

$$\delta : \text{Q} \times \Sigma \rightarrow \text{Q}.$$

The transition function may be extended to provide output as well as a state transition. Finite State machines may be represented by state diagrams. Each state accepts a finite number of inputs. The transition function provides rules that define the action of the machine for each input. Finite state machines may be deterministic or non-deterministic. A deterministic finite state machine changes to exactly one state for each input transition, while a non-deterministic automaton may have a choice of states to move to for a particular input.

Finite state automata can compute only very primitive functions and are therefore not an adequate model for computing. There are more powerful automata that compute whatever is computable. The Turing machine is essentially a finite automaton with an infinite storage (memory), and anything that is computable is computable by a Turing machine. The memory of the Turing machine is a tape which consists of an infinite number of one-dimensional cells. The Turing machine provides a mathematical abstraction of computer execution and storage, as well as providing a mathematical definition of an algorithm.

4.5.11 The Parnas Way

Parnas has been influential in the computing field, and his ideas on the specification, design, implementation, maintenance, and documentation of computer software remain important. He advocates a solid engineering approach to the development of high-quality software and argues that the role of the engineer is to apply scientific principles and mathematics to design and develop useful products. He argues that computer scientists should be educated as engineers and provided with the right scientific and mathematical background to do their work effectively. His contributions to software engineering include:

- Tabular expressions
  Tabular expressions are mathematical tables for specifying requirements and are also used in design. They enable complex predicate logic expressions to be represented in a simpler form.
- Mathematical documentation
  Parnas advocates the use of mathematical documents for software engineering that are precise and complete.
- Requirements specification
  He advocates the use of mathematical relations to specify the requirements precisely.
- Software design
  His contribution to software design includes information hiding that allows software to be designed for change. A module is characterised by its knowledge of a design decision (secret) that it hides from all others. Every information-hiding module has an interface that provides the only means to access the services provided by the modules. The interface hides the module's implementation. Information hiding[30] is used in object-oriented programming.

[30] I find it surprising that many in the object-oriented world seem unaware that information hiding goes back to the early 1970s and many have never heard of Parnas.

- Software inspections
  His approach to software inspections is quite distinct from the popular Fagan or Gilb inspection methodologies. The reviewers are required to take an active part in the inspection and they are provided with a list of questions by the author. The reviewers are required to provide documentation of their analysis to justify the answers to the individual questions. The inspections involve the production of mathematical tables, and may be applied to the actual software or documents.
- Predicate logic
  Parnas has introduced an approach to deal with undefined values in predicate logic expressions. The approach preserves the two-valued logic, and is quite distinct from the logic of partial functions developed by Cliff Jones. The latter is a three-valued logic.

4.5.12 Unified Modeling Language

The unified modeling language (UML) is a visual modeling language for software systems, and it facilitates the understanding of the architecture of the system, and in managing the complexity of large systems. It was developed by Jim Rumbaugh, Grady Booch, and Ivar Jacobson [Jac:04] as a notation for modeling object-oriented systems.

UML allows the same information to be presented in many different ways, and there are several UML diagrams providing different viewpoints of the system. Use cases describe scenarios or sequences of actions for the system from the user's viewpoint. A simple example is the operation of an ATM machine. Typical user operations at an ATM machine include the balance inquiry operation, the withdrawal of cash, and the transfer of funds from one account to another. These operations can be described with UML use-case diagrams.

Class and object diagrams are a part of UML and the object diagram is related to the class diagram in that the object is an instance of the class. There will generally be several objects associated with the class. The class diagram describes the data structure and the allowed operations on the data structure. The concept of class and objects are taken from object-oriented design. Two key classes are customers and accounts for an ATM system, and this includes the data structure for customers and accounts, and also the operations on customers and accounts. The operations include adding or removing a customer and operations to debit or credit an account. The objects of the class are the actual customers of the bank and their corresponding accounts.

Sequence diagrams show the interaction between objects/classes in the system for each use case. The sequences of interactions between objects for an ATM operation to check the balance of an account is illustrated in a sequence diagram that illustrates:

- Customer inserts the card into the ATM machine.
- PIN number is requested by the ATM machine.

- The customer then enters the PIN number.
- The ATM machine contacts the bank for verification of the number.
- The bank confirms the validity of the number and the customer then selects the balance inquiry.
- The ATM contacts the bank to request the balance of the particular account and the bank sends the details to the ATM machine.
- The balance is displayed on the screen of the ATM machine.
- The customer then withdraws the card.

UML activity diagrams are similar to flow charts. They are used to show the sequence of activities in a use case and include the specification of decision branches and parallel activities. The sequence of activities for the ATM operation to check the balance of an account may be shown in an activity diagram that illustrates:

- Card insertion
- Wait for PIN to be entered.
- Validate PIN.
- If Valid then check balance on account and Display balance.
- Otherwise return to 1.

State diagrams (or state charts) show the dynamic behaviour of a class and how different operations result in a change of state. There is an initial state and a final state, and the different operations result in different states being entered and exited.

There are several other UML diagrams including the collaboration diagram which is similar to the sequence diagram except that the sequencing is shown via a number system. UML offers a rich notation to model software systems and to understand the proposed system from different viewpoints. The main advantages of UML are listed in Table 4.4.

There is more detailed information on UML in [ORg:06].

4.5.12.1 Miscellaneous Specification Languages

The RAISE (Rigorous Approach to Industrial software Engineering) project was a European ESPRIT-funded project. Its objective [Geo:91] was to produce a method

Table 4.4 Advantages of UML

Advantages of UML
State of the art visual modeling language with a rich expressive notation.
Study of the proposed system before implementation
Visualization of architecture design of the system.
Mechanism to manage complexity of a large system.
Visualization of system from different viewpoints.
Enhanced understanding of implications of user behavior.
Use cases allow description of typical user behavior.
A mechanism to communicate the proposed behaviour of the software system. This describes what it will do and what to test against.

for the rigorous development of software, based on a wide-spectrum specification language, and accompanied by tool support. It considered standard VDM to be deficient, in that it lacked modularity, and as it was unable to deal with concurrency. The RAISE specification language (RSL) was designed to address these deficiencies, and an algebraic approach is adopted. Comprehensive support is available from the RAISE tools.

The RAISE method (as distinct from its specification language) covers the software lifecycle, from requirements analysis to code generation. This is achieved via a number of design steps, in which the specification is gradually made more concrete, until ultimately a specification that may be transferred into code is reached. The RAISE toolset includes library tools for storing and retrieving modules, and translators from subsets of RSL into Ada and C++.

The Specification and Descriptive Language (SDL) was developed to allow the behavior of telecommunication systems to be described and specified. It may be used at several levels of abstraction, ranging from a very broad overview of a system to detailed design. The behavior of the system is considered as the combined behavior of the processes in the system, and the latter is considered to be an extended finite state machine, i.e., a finite state machine that can use and manipulate data stored in variables local to the machine. Processes may cooperate via signals (i.e., discrete messages) and exhibit deterministic behavior.

A graphical language is employed to describe processes and this involves graphical representation of states, input, output, and decisions. Channels enable communication between blocks (containing processes) and the system (containing blocks connected by channels) and its environment. SDL supports time constraints via the timer construct. The graphical language has a corresponding equivalent textual representation.

4.5.13 Proof and Formal Methods

The word *proof* has several connotations in various disciplines; for example, in a court of law, the defendant is assumed innocent until proven guilty. The proof of the guilt of the defendant may take the form of certain facts in relation to the movements of the defendant, the defendant's circumstances, the defendant's alibi, statements taken from witnesses, rebuttal arguments from the defense, and certain theories produced by the prosecution or defense. Ultimately, in the case of a trial by jury, the defendant is judged guilty or not guilty depending on the extent to which the jury has been convinced by the arguments made by the prosecution and defense.

A mathematical proof typically includes natural language and mathematical symbols, and often many of the tedious details of the proof are omitted. The strategy of proof in proving a conjecture tends to be *a divide and conquer* technique; i.e., breaking the conjecture down into subgoals and then attempting to prove the subgoals. Most proofs in formal methods are concerned with cross-checking on the details of the specification or are concerned with checking the validity of refinement steps, or proofs that certain properties are satisfied by the specification. There are often many tedious lemmas to be proved, and theorem provers[31] are essential in assisting with this. Machine proof needs to be explicit, and reliance on some brilliant insight is avoided. Proofs by hand are notorious for containing errors or jumps in reasoning, as discussed in chapter one of [HB:95], while machine proofs are often extremely lengthy and unreadable. They generally help to avoid errors and jumps in reasoning, as every step in the proof needs to be justified.

A mathematical proof consists of a sequence of formulae, where each element is either an axiom or derived from a previous element in the series by applying a fixed set of mechanical rules. One well-known theorem prover is the Boyer/Moore theorem prover [BoM:85]. There is an interesting case in the literature concerning the proof of correctness of the VIPER microprocessor[32] [Tie:91], and the actual machine proof consisted of several million formulae.

Theorem provers are invaluable in resolving many of the thousands of proof obligations that arise from a formal specification, and it is not feasible to apply formal methods in an industrial environment without the use of machine assisted proof. Automated theorem proving is difficult, as often mathematicians prove a theorem with an initial intuitive feeling that the theorem is true. Human intervention to provide guidance or intuition improves the effectiveness of the theorem prover.

The proof of various properties about the programs increases confidence in the correctness of the program. However, an absolute proof of correctness is unlikely except for the most trivial of programs. A program may consist of legacy software which is assumed to work, or be created by compilers which are assumed to work. Theorem provers are programs which are assumed to function correctly. Therefore, in order to be absolutely certain one would also need to verify the hardware, customized-off-the-shelf software, subcontractor software, compilers, legacy software, the theorem prover itself, and every single execution path that the software system will be used for. The best that formal methods can claim is increased confidence in correctness of the software, rather than an absolute proof of correctness.

4.6 Software Inspections and Testing

Software inspections and testing play a key role in building quality into software products and verifying that the products are of high quality. The Fagan Inspection Methodology is a well-known software inspection methodology developed by Michael Fagan of IBM [Fag:76]. It is a seven-step process that identifies and removes errors in work products. There is a strong economic case for identifying defects as early as possible, as the cost of correction of increases the later a defect is discovered in the lifecycle. The Fagan inspection process mandates that requirement

[31] Most existing theorem provers are difficult to use and are for specialist use only. There is a need to improve the usability of theorem provers.

[32] As discussed earlier this verification was controversial with RSRE and Charter overselling VIPER as a chip design that conforms to its formal specification.

documents, design documents, source code, and test plans are all formally inspected by experts independent of the author of the deliverable to ensure quality.

There are various *roles* defined in the inspection process including the *moderator* who chairs the inspection. The moderator is skilled in the inspection process, and is responsible for ensuring that all of the participants receive the appropriate materials for the inspection, and that sufficient preparation has been done by all of the participants. The moderator will ensure that any major or minor errors identified are recorded, and that the speed of the inspection does not exceed the recommended guidelines. The *reader's* responsibility is to read or paraphrase the particular deliverable, and the *author* is the creator of the deliverable and has a special interest in ensuring that it is correct. The *tester* role is concerned with the test viewpoint.

The inspection process will consider whether a design is correct with respect to the requirements, and whether the source code is correct with respect to the design. There are seven stages in the inspection process and these are described in detail in [ORg:02]:

- Planning
- Overview
- Prepare
- Inspect
- Process improvement
- Re-work
- Follow-up

The errors identified in an inspection are classified into various types as defined by the Fagan methodology. A mature organization will record the inspection data in a database and this will enable analysis to be performed on the most common types of errors. The analysis will yield actions to be performed to minimize the re-occurrence of the most common defect types. Also, the data will enable the effectiveness of the organization in identifying errors in phase and detecting defects out of phase to be determined and improved. Another approach to software inspection has been defined by Tom Gilb [Glb:94]. Some organizations use less formal inspection methodologies such as pass-around peer reviews. This involves sending a deliverable to the reviewers and requesting comments by a certain date. The author is then responsible for collating the comments and acting upon them appropriately. The disadvantage of this approach is that control of the inspection is given to the author rather than an independent moderator. Another approach to peer reviews is available with the Prince 2 project methodology.

Software testing plays a key role in verifying that a software product is of high quality and conforms to the customer's quality expectations. Testing is both a constructive activity in that it is verifying the correctness of functionality, and it may be a destructive activity in that the objective is to find as many defects as possible in the software. The testing verifies that the requirements are correctly implemented as well as identifying whether any defects are present in the software product.

There are various types of testing such as unit testing, integration testing, system testing, performance testing, usability testing, regression testing, and customer acceptance testing. It needs to be planned to ensure that it is effective. Test cases will need to be prepared and executed, the results reported and any issues corrected and re-tested. The test cases will need to be appropriate to verify the correctness of the software.

The quality of the testing is dependent on the maturity of the test process, and a good test process will include:

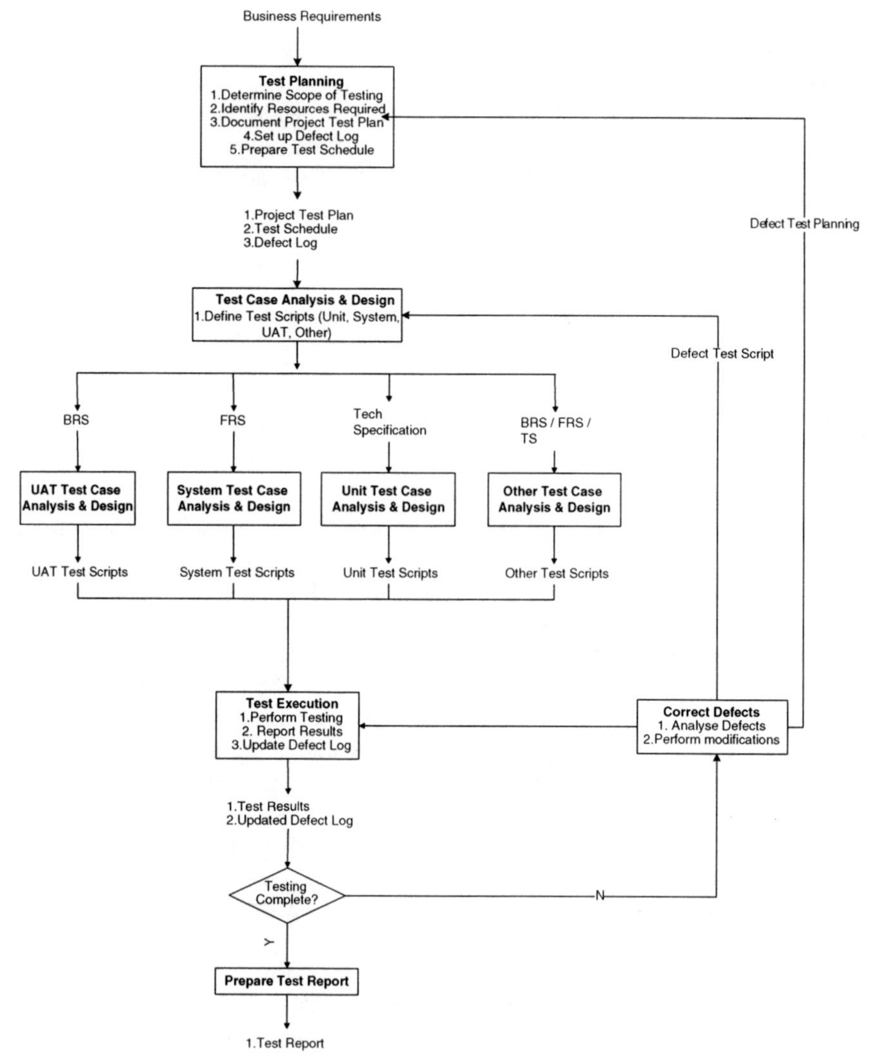

Fig. 4.10 Simplified test process

- Test planning and risk management
- Dedicated test environment and test tools
- Test case definition
- Test automation
- Formality in handover to test department
- Test execution
- Test result analysis
- Test reporting
- Measurements of test effectiveness
- Post mortem and test process improvement.

A simplified test process is sketched in Fig. 4.10:

Metrics are generally maintained to provide visibility into the effectiveness of the testing process. Testing is described in more detail in [ORg:02].

4.7 Process Maturity Models

The Software Engineering Institute (SEI) developed the Capability Maturity Model (CMM) in the early 1990s as a framework to help software organizations to improve their software process maturity, and to implement best practice in software and systems engineering. The SEI and many quality experts believe that there is a close relationship between the maturity of software processes and the quality of the delivered software product. The CMM built upon the work of quality gurus such as Deming [Dem:86] (Fig. 4.11), Juran [Jur:2000] (Fig. 4.12) and Crosby [Crs:79] whose ideas had been very effective in transforming manufacturing companies to produce high quality products and in reducing the costs associated with reworking defective products.

They recognized a need to focus on the process, and software organisations need to focus on the software development process as well as the product.

Early work on the CMM was done by Watts Humphrey at IBM [Hum:89] (Fig. 4.13). Humphrey moved to the SEI in the late 1980s, and worked with the other SEI experts to produce the first version of the CMM in 1991. The model has been developed further in recent years and it is now called the Capability Maturity Model Integration (CMMI®) [CKS:03].

The CMMI consists of five maturity levels with each maturity level (except level one) consisting of several process areas. Each process area consists of a set of goals that must be satisfied for the process area to be satisfied. The goals for the process area are implemented by practices related to that process area, and the implementation of these practices lead to an effective process. Processes need to be defined and documented. The users of the process need to receive appropriate training to enable them to carry out the process, and processes need to be enforced by independent audits.

The emphasis on level two of the CMMI is on maturing management practices such as project management, requirements management, configuration

Fig. 4.11 W. Edwards Deming
Courtesy of W. Edwards Deming Institute.

management, and so on. The emphasis on level three of the CMMI is to mature engineering and organisation practices. This maturity level includes peer reviews and testing, requirements development, software design and implementation practices, and so on. Level four is concerned with ensuring that key processes are performing within strict quantitative limits, and adjusting processes, where necessary, to perform within these defined limits. Level five is concerned with continuous process improvement which is quantitatively verified.

Maturity levels may not be skipped in the staged implementation of the CMMI. There is also a continuous representation of the CMMI which allows the organization to focus on improvements to key processes. However, in practice it is often necessary to implement several of the level two process areas before serious work can be done on implementing a process at a higher maturity level. The use of metrics [Fen:95, Glb:76] becomes more important as an organization matures, as metrics allow the performance of an organization to be objectively judged. The higher CMMI maturity levels set quantitative levels for processes to perform within.

The CMMI allows organizations to benchmark themselves against other similar organizations. This is done by formal SEI approved SCAMPI appraisals conducted

Fig. 4.12 W. Joseph Juran
Courtesy of Juran Institute.

Fig. 4.13 Watts Humphrey
Courtesy of Watts Humphrey.

by an authorized SCAMPI lead appraiser. The results of a SCAMPI appraisal are generally reported back to the SEI, and there is a strict qualification process to become an authorized SCAMPI lead appraiser. An appraisal is useful in verifying that an organization has improved, and it enables the organisation to prioritise improvements for the next improvement cycle.

The time required to implement the CMMI in an organization depends on the current maturity and size of the organization. It generally takes 1–2 years to implement maturity level 2, and a further 1–2 years to implement level 3.

4.8 Review Questions

1. Describe the crisis in software in the 1960s and the birth of software engineering.
2. Describe waterfall and spiral lifecycle models including their advantages and disadvantages.
3. Discuss Floyd's contribution to software engineering and how it led to Hoare's axiomatic semantics.
4. Describe the mathematics that is potentially useful in software engineering.
5. Describe formal methods and its applications to software engineering. Explain when their use should be considered in software engineering.
6. Discuss any tools to support formal methods that you are familiar with.
7. Discuss the similarities and differences between Z and VDM.
8. Discuss the similaritities and differences between the model-oriented approach and the axiomatic approach to formal methods.
9. Discuss UML and its applicability to software engineering.
10. Discuss the applicability of software inspections and testing to software engineering.
11. Discuss the Capability Maturity Model and its applicability to software engineering.

4.9 Summary

This chapter discussed the birth of software engineering from the "software crisis" that existed in the 1960s. The computer scientists that attended the first software engineering conference at Garmisch in Germany recognised the need for sound methodologies to design, develop and maintain software to meet customer needs.

Classical engineering has a successful track record in building high-quality products that are safe to use. It is therefore natural to consider using an engineering approach to developing software, and this involves identifying the customer requirements, carrying out a rigorous design to meet the requirements, developing and coding a solution to meet the design, and conducting appropriate inspections and testing to verify the correctness of the solution.

Mathematics plays a key role in engineering to assist with design and verification of software products. It is therefore reasonable to apply appropriate mathematics in software engineering (especially for safety critical systems) to assure that the delivered systems conform to the requirements. The extent to which mathematics will need to be used will vary, and in many cases peer reviews and testing will be sufficient to build quality into the software product. In other cases, and especially with safety and security critical applications, it is desirable to have the extra assurance that may be provided by mathematical techniques.

Various mathematical approaches were considered including Z, VDM and CSP. The propositional calculus and predicate calculus were discussed as well as the nature of proof and the importance of tools (and theorem provers).

The unified modeling language (UML) has become popular in recent years. It is a visual modeling language for software systems and facilitates an understanding of the architecture of the system, and in managing the complexity of large systems.

There is a lot more to the successful delivery of a project than just the use of mathematics or peer reviews and testing. Sound project management and quality management practices are essential, as a project that is not properly managed will suffer from schedule, budget or cost overruns as well as problems with quality.

Maturity models such as the CMMI can assist organisations in maturing key management and engineering practices that are essential for the successful delivery of high-quality software. The use of the CMMI helps companies in their goals to deliver high-quality software systems that are consistently on time and consistently meet business requirements.

Chapter 5
Artificial Intelligence and Expert Systems

Key Topics

Turing Test
Searle's Chinese Room
  Philosophical Problems in AI
  Cognitive Psychology
  Linguistics
  Logic and AI
  Robots
  Cybernetics
  Neural Networks
  Expert Systems

5.1 Introduction

The ultimate goal of Artificial Intelligence is to create a thinking machine that is intelligent, has consciousness, has the ability to learn, has free will, and is ethical. The field involves several disciplines such as philosophy, psychology, linguistics, machine vision, cognitive science, mathematics, logic and ethics. Artificial Intelligence is a young field and the term was coined by John McCarthy and others in 1956. Alan Turing had earlier devised the Turing Test as a way to test the intelligent behaviour of a machine. There are deep philosophical problems in Artificial Intelligence, and some researchers believe that its goals are impossible or incoherent. These views are shared by Hubert Dreyfus and John Searle. Even if Artificial Intelligence is possible there are moral issues to consider such as the exploitation of artificial machines by humans, and whether it is ethical to do this. Weizembaum[1] has argued that Artificial Intelligence is unethical.

[1] Weizenbaum was a psychologist who invented the ELIZA program. This program simulated a psychologist in dialogue with a patient. He was initially an advocate of Artificial Intelligence but later became a critic.

Perhaps, one of the earliest references to creating life by artificial means is that of the classical myth of Pygmalion. Pygmalion was a sculptor who carved a woman out of ivory. The sculpture was so realistic that he fell in love with it, and offered the statue presents and prayed to Aphrodite the goddess of love. Aphrodite took pity on him and brought the statue (Galathea) to life. Pygmalion and Galathea married and had a son Paphos.

Aristotle developed syllogistic logic in the fourth century B.C. This was the first formal deductive reasoning system, and was discussed in Chapter 1. A syllogism consists of two premises (or propositions) and one conclusion. Each premise consists of two terms and there is one common middle term. The conclusion consists of the two unrelated terms from the premises. This is best illustrated by an example:

Premise 1	All Greeks are Mortal
Premise 2	Socrates is a Greek.
	....
Conclusion	Socrates is Mortal

Aristotle studied the various syllogisms and determined which were valid or invalid. Aristotle's syllogistic logic was used widely until the nineteenth century. It allowed further truths to be derived from existing truths.

Ramon Llull was a medieval writer and philosopher in the thirteenth century. He was born in Palma, Mallorca in the Balearic Islands. He did some pioneering work in computation theory by designing a method of combining attributes selected from a number of lists. He intended his method to be used as a debating tool to convert Muslims to the Christian faith through logic and reason. Llull's machine allowed a reader to enter an argument or question about the Christian faith, and the reader would then turn to the appropriate index and page to find the correct answer.

Llull's method was an early attempt to use logical means to produce knowledge. He hoped to show that Christian doctrines could be obtained artificially from a fixed set of preliminary ideas or undeniable truths. For example, the attributes of God as a supremely good, benevolent, omnipotent being were undeniable truths, and agreed between the main mono-theistic faiths of Islam, Judaism and Christianity. He believed that each field of knowledge had a limited number of basic undeniable truths, and that everything could be understood about the fields of knowledge by studying combinations of these elementary truths.

Llull's machine is called the Lullian Circle and it consists of two or more paper discs inscribed with alphabetic letters or symbols. These discs may be rotated separately to generate a large combination of ideas. A number of terms were laid around the full circumference of the circle, and repeated on an inner circle that could be rotated. These combinations were said to show all possible truths about the subject of the circle. Llull was opposed by the Grand Inquisitor of Aragon, and some of his writings were banned by the pope.

The creation of man by God is described in Genesis in the Bible. There are several stories of attempts by man to create life from inanimate objects: for example, the creation of the monster in Mary Shelly's Frankenstein. These stories go back to an earlier period: for example, there are stories of the creation of the golem in

Prague dating back to sixteenth century. The word "*golem*" is used in the Bible to refer to an incomplete substance, and the word is generally used today to describe someone who is clumsy or slow. All golems were created from mud by a very holy person who was very close to God, and who was able to create life. However, the life that the holy person could create would always be a shadow of one created by God. Golems became servants to the holy men but were unable to speak. The possession of a golem servant was a sign of the wisdom and holiness of the beholder. Other attributes of the golem were added over time, and included inscription with magical words to keep the golem animated. This also allowed the golem to be deactivated by inscribing the word "death" on the forehead of the golem.

The most famous golem story involved Rabbi Judah Loew of Prague. He was a sixteenth century rabbi and is said to have created a golem to defend the Jews of Prague from attack. According to the story the golem was created from clay on the banks of the Vltava river in Prague, and was brought to life by the Rabbi by the recitation of special incantations in Hebrew. However, as the golem grew bigger he became violent and started killing people in Prague. The Rabbi agreed to destroy the golem if the anti-semitic violence against the Jews of Prague stopped. The golem was destroyed by erasing the first letter of the word "*emet*" from the golem's forehead to make the Hebrew word "*met*", meaning death. The remains of Prague's golem are believed to be in Prague. Golems were not very bright as they were created by man and not God. Whenever, a golem was commanded to perform a task, it did so by following the instructions literally.

The golem has similarities with Mary Shelley's Frankenstein. Frankstein's monster is created by an overambitious scientist who is punished for his blasphemy of creation (in that creation is for God alone). The monster feels rejected following creation, and inflicts a horrible revenge on its creator. The story of the golum was given a more modern version in the Czech play "Rossums Universal Robots". This science fiction play by Capek appeared in Prague in 1921. It was translated into English and appeared in London in 1923. It contains the first reference to the term "robot", and the play considers the exploitation of artificial workers in a factory. The robots (or androids) are initially happy to serve humans, but become unhappy with their existence over a period of time. The fundamental question that the play is considering is whether the robots are being exploited, and if so, is this ethical, and how should the robots respond to their exploitation. It eventually leads to a revolt by the robots and the extermination of the human race. Capek considered the exploitation of non-human forms of life (the newts) in his play "War with the Newts" which was a satire on Europe and the world in the 1930s.

5.2 Descartes

Rene Descartes (Fig. 5.1) was an influential French mathematician, scientist and philosopher. He was born in a village in the Loire valley in France in 1596, and studied law at the University of Poitiers. He never practiced as a lawyer and instead served Prince Maurice of Nassau in the Netherlands. He invented the Cartesian coordinate system that is used in plane geometry and algebra. In this system, each

Fig. 5.1 Rene Descartes

point on the plane is identified through two numbers: the x-coordinate and the y-coordinate.

Descartes made important contributions to philosophy and he attempted to develop a fundamental set of principles which can be known to be true. He employed scepticism to arrive at this core set of truths, and his approach was to renounce any idea that could be doubted. He rejected the senses since the senses can deceive, and therefore they are not a sound source of knowledge. For example, during a dream the subject perceives stimuli that appear to be real, but these have no existence outside the subject's mind. Therefore, it is inappropriate to rely on one's senses as the foundation of knowledge. He argued that a powerful "evil demon or mad-scientist" could exist who sets out to manipulate and deceive subjects, thereby preventing them from knowing the true nature of reality. The evil demon could bring the subject into existence including an implanted memory. The question is how can one know for certain what is true given the limitations of the senses.

From this, Descartes deduced that there was one single principle that must be true. He argued that even if the is being deceived, then clearly he is thinking and must exist. This principle of existence or being is more famously known as "*cogito, ergo sum*" (I think, therefore I am). Descartes argued that this existence can be applied to the present only, as memory may be manipulated and therefore doubted. Further, the only existence that he sure of is that he is a "*thinking thing*". He cannot be sure of the existence of his body as his body is perceived by his senses which he has proven to be unreliable. Therefore, his mind or thinking thing is the only thing about him that cannot be doubted. His mind is used to make judgments, and to deal with un-reliable perceptions received via the senses.

Descartes constructed a system of knowledge from this one principle using the deductive method. He deduced the existence of a benevolent God using the ontological argument. This argument had first been formulated by St. Anselm in the eleventh century, and the proof of God was based on the definition and perfection of God ("The Being that nothing greater can be conceived"). Descartes argument in the fifth meditation in his Mediations on First Philosophy [Des:99] is that we have an innate idea of a supremely perfect being (God), and that God's existence may be inferred immediately from the innate idea of a supremely perfect being. The argument may be summarised as follows:

1. I have an innate idea of a supremely perfect being (i.e. God).
2. Necessarily, existence is a perfection.
3. Therefore God exists.

He then argued that as God is benevolent that he can have some trust in the reality that his senses provide, as God has provided him with a thinking mind and does not wish to deceive him. He then argued that knowledge of the external world can be obtained by both perception and deduction. In other words, he argues that reason or rationalism is the only reliable method of obtaining knowledge. Descartes arguments provided an answer to the school of scepticism that held the view that knowledge, and knowledge of existence of the external world was impossible.

Descartes was a dualist and he makes a clear mind-body distinction.[2] He states that there are two substances in the universe: mental substances and bodily substances. His proof of the existence of God and the external world are controversial and have been criticised by the sceptics and others. One well known argument against Descartes is the "the brain in the vat" argument. This is a thought experiment in which a mad scientist removes a person's brain from their body and places it in a vat and connects its neurons by wires to a supercomputer. The computer provides the disembodied brain with the electrical impulses that the brain would normally receive. The computer could then simulate reality, and the disembodied brain would have conscious experiences and would receive the same impulses as if it were inside a person's skull. There is no way to tell whether the brain is inside the vat or inside a person. The perception of a "cat" in the first case reflects the reality. However, for the second case where the brain is in the vat the perception is false, and does not correspond to reality. Since it is impossible to know whether your brain is in a vat or inside your skull it is therefore impossible to know whether your belief is valid or not.

Descartes believed that the bodies of animals are complex living machines without feelings. He dissected many animals for experiments and this included

[2] The mind-body distinction is very relevant in AI. The human mind may be considered to be a piece of software which the human brain implements. Therefore, in principle, it should be possible to code this program on a von Neumann computer yielding a mental machine. Cognitive science was originally based on this paradigm.

vivisection (i.e., the cutting up of live animals). Vivisection has become controversial in recent years on ethical grounds, and the question is whether it right that animals should be used and subject to suffering to further human interests. This debate has centred on whether animals have rights being animals given that humans have inalienable rights being human. Others have argued against vivisection claiming that it is not necessary in scientific research. Some have argued for vivisection on utilitarian grounds, with others stating that animals have rights, and should not be used as means to an end, irrespective of the expected benefits to humans.

Descartes experiments led him to believe that the actions and behaviour of non-human animals can be fully accounted for by mechanistic means, and without reference to the operations of the mind. Descartes also realised from his experiments that a lot of human behaviour is like that of animals in that it has a mechanistic explanation. For example, physiological functions and blinking are human behaviours that have a mechanistic explanation. Descartes was of the view that well-designed automata[3] could mimic many parts of human behaviour. He argued that the key differentiators between human and animal behaviour were that humans could adapt to widely varying situations, and also had the ability to use language. The use of language illustrates the power of the use of thought, and it clearly differentiates humans from animals. Animals do not possess the ability to use language for communication or reason. This, he argues, provides evidence for the presence of a soul associated with the human body. In essence, animals are pure machines, whereas humans are machines with minds (or souls).

Descartes was strongly against attributing minds to animals as it would undermine religious belief. The ascribing of thought to animals involves giving them immortal souls, since Descartes considered mental substances to be indivisible. Descartes associated the concept of mind with the Christian concept of the soul, and he argued that if animals had minds (or souls) then man's hope for an afterlife could be no more than that of flies or bats. He argued that since animals did not have minds (or souls), then this meant that they did not suffer, and that man is absolved from guilt in slaughtering and eating animals.

However, if Descartes could not infer the existence of minds in animals from their behaviour, then how could he infer the existence of minds in other humans from their behavior? His analogy could just as easily be extended to man's mind as to his body as God is capable of contriving both human and animal automata.

The significance of Descartes in the field of Artificial Intelligence is that the Cartesian dualism that humans seem to possess would need to be reflected among artificial machines. Humans seem to have a distinct sense of "I" as distinct from the body, and the "I" seems to represent some core sense or essence of being that is unchanged throughout the person's life. It somehow represents personhood, as distinct from the physical characteristics of a person that are inherited genetically. The challenge for the AI community in the longer term is to construct a machine

---

[3] An automaton is a self-operating machine or mechanism that behaves and responds in a mechanical way.

that (in a sense) possesses Cartesian dualism. That is, the long-term[4] goal of AI is to produce a machine that has awareness of itself as well as its environment.

5.3 The Field of Artificial Intelligence

The foundations for Artificial Intelligence goes back to the foundations of computing, and it includes the work of famous scientists, engineers and mathematicians. These include names such as Boole, Babbage, Shannon, von Neumann and Turing. There is an excellent account of the various disciplines in the AI field, as well as key problems to be solved if the field is to progress in [ONu:95]. Artificial Intelligence is a multi-disciplinary field and its many branches include:

- Computing
- Logic and Philosophy
- Psychology
- Linguistics
- Machine Vision
- Computability
- Epistemology and Knowledge representation

The British mathematician, Alan Turing, contributed to the debate concerning thinking machines, consciousness and intelligence in the early 1950s [Tur:50]. He devised the famous "Turing Test" to judge whether a machine was conscious and intelligent. Turing's paper was very influential as it raised the idea of the possibility of programming a computer to behave intelligently. Shannon considered the problem of writing a chess program in the late 1940s, and distinguished between a brute force strategy where the program could look at every combination of moves or a strategy where knowledge of chess could be used to select and examine a subset of available moves. Turing also wrote programs for a computer to play chess, and the ability of a program to play chess is a skill that is considered intelligent, even though the machine itself is not conscious that it is playing chess.

The early computer chess programs were not very sophisticated and performed poorly against humans. The modern chess programs have been quite successful, and have advantages over humans in terms of computational speed in considering combinations of moves. The chess grandmaster, Kasparov, was defeated by the IBM chess program "Deep Blue" in a six game match in 1997. The computer and Kasparov were even going into the final game which Kasparov lost. Kasparov asked for a re-match against Deep Blue, but IBM decided instead to retire the machine. Kasparov also played a 4-game match against the chess playing program X3D Fritz in 2003. The result of this match was a draw.

---

[4] This long-term goal may be hundreds of years as there is unlikely to be an early breakthrough in machine intelligence as there are deep philosophical problems to be solved. It took the human species hundreds of thousands of years to evolve to its current levels of intelligence.

The origin of the term "Artificial Intelligence" is in work done on the proposal for Dartmouth Summer Research Project on Artificial Intelligence. This proposal was written by John McCarthy (Fig. 5.2) and others in 1955, and the research project took place in the summer of 1956. Herbert A. Simon and Alan Newell did work in 1956 [NeS:56] on a program called "Logic Theorist" or "LT". This program was the world's first theorem prover, and could independently provide proofs of various theorems[5] in Russell's and Whitehead's Principia Mathematica [RuW:10]. The proof of a theorem requires creativity and intelligence, and the LT program demonstrated that computers had the ability to perform intelligent operations. Simon's and Newell's LT program was demonstrated at the Dartmouth summer conference on Artificial Intelligence. This program was very influential as it demonstrated that knowledge and information could be programmed into a computer, and that a computer had the ability to solve theorems in mathematics.

John McCarthy's approach to knowledge representation is to employ logic as the method of representing information in computer memory. He proposed a program called the Advice Taker in his influential paper "Programs with Common Sense" [Mc:59]. The idea was that this program would be able to draw conclusions from a set of premises, and McCarthy states that a program has common sense if it is capable of automatically deducing for itself "a sufficiently wide class of immediate consequences of anything it is told and what it already knows".

The Advice Taker aims to use logic to represent knowledge (i.e., premises that are taken to be true), and it then applies the deductive method to deduce further truths from the relevant premises.[6] That is, the program manipulates the formal language (e.g., predicate logic), and provides a conclusion that may be a statement or an imperative. McCarthy's Advice Taker differs from Simon's and Newell's LT program in that the latter program is a heuristic for deduction. McCarthy envisaged that the Advice Taker would be a program that would be able to learn and improve. This would be done by making statements to the program, and telling it about its symbolic environment. The program will have available to it all the logical consequences of what it has already been told and previous knowledge. McCarthy's desire was to create programs to learn from their experience as effectively as humans do.

McCarthy's philosophy is that common sense knowledge and reasoning can be formalised with logic. A particular system is described by a set of sentences in logic. These logic sentences represent all that is known about the world in general, and what is known about the particular situation and the goals of the systems. The

---

[5] Russell is reputed to have remarked that he was delighted to see that the Principia Mathematica could be done by machine, and that if he and Whitehead had known this in advance that they would not have wasted 10 years doing this work by hand in the early twentieth century. The LT program succeeded in proving 38 of the 52 theorems in Chapter of Principia Mathematica. Its approach was to start with the theorem to be proved and to then search for relevant axioms and operators to prove the theorem.

[6] Of course, the machine would somehow need to know what premises are relevant and should be selected for to apply the deductive method from the many premises that are already encoded.

Fig. 5.2 John McCarthy
Courtesy of John McCarthy.

program then performs actions that it infers are appropriate for achieving its goals. That is, common sense[7] knowledge is formalised by logic, and common sense problems are solved by logical reasoning.

5.3.1 Turing Test and Strong AI

Turing contributed to the debate concerning artificial intelligence in his 1950 paper on Computing, machinery and intelligence [Tur:50]. Turing's paper considered whether it could be possible for a machine to be conscious and to think. He predicted that it would be possible to speak of machines thinking and he devised a famous experiment that would allow a computer to be judged as a conscious and thinking machine. This is known as the "Turing Test". The test itself is an adaptation of a well-known party game which involves three participants. One of them, the judge, is placed in a separate room from the other two: one is a male and the other is a female. Questions and responses are typed and passed under the door. The objective

[7] Common sense includes basic facts about events, beliefs, actions, knowledge and desires. It also includes basic facts about objects and their properties.

of the game is for the judge to determine which participant is male and which is female. The male is allowed to deceive the judge whereas the female is supposed to assist.

Turing adapted this game by allowing the role of the male to be played by a computer. If the judge could not tell which of the two participants was human or machine, then the computer could be considered intelligent. The test involves a judge who is engaged in a natural language conversation with two other parties, one party is a human and the other is a machine. If the judge cannot determine which is machine and which is human, then the machine is said to have passed the "Turing Test". That is, a machine that passes the Turing Test must be considered intelligent, as it is indistinguishable from a human. The test is applied to test the linguistic capability of the machine rather than the audio capability, and the conversation is limited to a text only channel.

Turing's work on "thinking machines" caused a great deal of public controversy as defenders of traditional values attacked the idea of machine intelligence. Turing's paper led to a debate concerning the nature of intelligence. There has been no machine developed, to date, that has passed the Turing test.

Turing strongly believed that machines would eventually be developed that would stand a good chance of passing the "Turing Test". He considered the operation of "thought" to be equivalent to the operation of a discrete state machine. Such a machine may be simulated by a program that runs on a single, universal machine, i.e. a computer.

Turing viewpoint that a machine will one day pass the Turing Test and be considered intelligent is known as "*strong artificial intelligence*". It states that a computer with the right program would have the mental properties of humans. There are a number of objections to strong AI, and one well-known rebuttal is that of Searle's Chinese Room argument [Sea:80].

The Chinese Room argument is a thought experiment that is used to demonstrate that a machine will never have the same cognitive qualities as a human. A man is placed into a room into which Chinese writing symbols are given to him. He has the knowledge of what symbol to use to respond to each symbol that is presented to him. However, he has no idea as to what each symbol means. Essentially, he is communicating with the person who is giving the symbols to him, and answering any questions that are posed, without the slightest understanding of what he is doing and what the symbols mean.

The closed room has two slots where slot 1 is used to input the Chinese characters. The person in the room has no understanding of what these Chinese characters mean. However, a rulebook is provided that allows new Chinese character to be created from the Chinese characters which have already been input. Slot 2 is used to output the Chinese characters. There are essentially three steps in the process:

1. Chinese characters are entered through slot 1.
2. The rulebook is employed to construct new Chinese characters.
3. Chinese characters are outputted to slot 2.

The process is essentially that of a computer program which has an input; performs a computation based on the input; and then finally produces an output. Further, the rulebook is such that people outside the room are able to send questions such as "How are you?". The responses to these questions are provided following the rulebook, and meaningful answers are provided. That is, the computer program simulates a human being who understands Chinese, even though the person has not the slightest understanding of the language.

The question "Do you understand Chinese?" could potentially be asked, and the rulebook would be consulted to produce the answer "Yes, of course" that is, despite of the fact that the person inside the room has not the faintest idea of what is going on, it will appear to the person outside the room that the person inside is knowledgeable on Chinese. The person in the room is just following rules without understanding.

Searle has essentially constructed a machine which can never be mental. Changing the program essentially means changing the rulebook, and this does not increase understanding. The strong artificial intelligence thesis states that given the right program, *any* machine running it would be mental. However, Searle argues that the program for this Chinese room would not understand anything, and that therefore the strong AI thesis must be false. In other words, Searle's Chinese room argument is a rebuttal of strong AI by showing that a program running on a machine that appears to be intelligent has no understanding whatsoever of the symbols that it is manipulating. That is, given any rulebook (i.e., program), the person would never understand the meanings of those characters that are manipulated.

Searle argument essentially states that just because the machine acts like it knows what is going on, it only knows what it is programmed to know. It differs from humans in that it is not aware of the situation like humans are. The question is whether a machine can only do what it is told to do, or whether it may think for itself and have consciousness. Searle's argument suggests that machines may not have intelligence or consciousness, and the Chinese room argument applies to any Turing equivalent computer simulation.

There are several rebuttals of Searle's position[8] and one well-known rebuttal attempt is the "System Reply" argument. This reply argues that if a result associated with intelligence is produced, then intelligence must be found somewhere in the system. The proponents of this argument draw an analogy between the human brain and its constituents. None of its constituents have intelligence but the system as a whole (i.e., the brain) exhibits intelligence. Similarly, the parts of the Chinese room may lack intelligence, but the system as a whole is intelligence. However, this rebuttal argument has been criticized by others as begging the question.

---

[8] I don't believe that Searle's argument proves that Strong AI is impossible. However, I am not expecting to see intelligent machines anytime soon.

5.4 Philosophy and AI

Artificial Intelligence includes the study of knowledge and the mind, and there are deep philosophical problems to be solved in the field. The philosophy of epistemology and the philosophy of mind are concerned with fundamental questions such as:

- What is being?
- What is knowledge?
- What is mind?
- What is consciousness?

Early work on philosophy was done by the Greeks as they attempted to understand the world and the nature of being and reality.[9] Thales and the Miletians[10] attempted to find an underlying principle: e.g., water, or for other Miletians, earth, air, wind and fire, where such a principle that would explain everything. Pythagoras believed that mathematical numbers were the basic underlying principle, and that everything (e.g., music) could be explained in terms of number. Plato distinguished between the world of appearances and the world of reality. He argued that the world of appearances resembles the flickering shadows on a cave wall, whereas reality is in the world of ideas[11] or forms, in which objects of this world somehow participate. Aristotle wondered how many forms there are: for example, is there a separate form for a dog, breed of dog, etc. Aristotle proposed the framework of a substance which includes form plus matter. For example, the matter of a wooden chair is the wood that it is composed of, and its form is the general form of a chair. The problem of being is relevant to the mind-body problem in AI.

Acquinus was a medieval scholastic philosopher who was deeply influenced by Aristotle.[12] Acquinus distinguished between matter and form and also considered questions such as existence and understanding. Modern Thomists,[13] such as Bernart Lonergan have developed his ideas into the nature of understanding and cognition. Lonergans's study of cognition [Lon:58] was influenced by the eureka step in the discovery of Archimedes, and also the in the sudden insight in which the solution of a mathematical puzzle comes about. Lonergan outlines the steps associated with the sudden insight that comes in solving a mathematical puzzle or scientific discovery. Artificial Intelligence requires a theory of cognition if it is to make serious progress.

---

[9] The study of philosophical questions related to being is referred to as Ontology (or Metaphysics). The study of philosophical questions related to knowledge is referred to as epistemology.

[10] The term "Miletians" refers to inhabitants of the Greek city state Miletus which is located in modern Turkey. Anaximander and Anaximenes were two other Miletians who made contributions to early Greek philosophy approx 600 BC.

[11] Plato was an Idealist: i.e., that reality is in the world of ideas rather than the external world. Realists (in contrast) believe that the external world corresponds to our mental ideas.

[12] He refers to Aristotle as "The Philosopher" in Sumna Theologicae.

[13] The term "Thomist" denotes a follower of the philosophy of St. Thomas Acquinus.

Descartes was discussed earlier and his influence on the philosophy of mind and AI is significant. He was deeply concerned with a firm foundation for knowledge and being, and his approach was to doubt everything that could be doubted except his own existence. He deduced his own existence from the fact that he could think: i.e., *Cogito, ergo sum.*[14] This thinking thing (*res cogitans* or mind/soul) was distinct from the rest of nature and interacted with the world through the senses to gain knowledge. Knowledge was gained by mental operations using the deductive method, where starting from the premises that are known to be true, further truths could be logically deduced. Descartes founded what would become known as the Rationalist school of philosophy where knowledge was derived solely by human reasoning. He distinguished between the mind and the body (Cartesian dualism), and the analogy of the mind in AI would be the central processor unit (or an AI program) of a computer, with knowledge gained by sense perception by the computer hardware (e.g., machine vision) and logical deduction.

British Empiricism rejected the Rationalist position, and stressed the importance of empirical data in gaining knowledge about the world. Its philosophy argued that all knowledge is derived from sense experience. It consisted of philosophers such as Hobbes, Locke, Berkeley[15] and Hume. Locke argued that a child's mind is a blank slate (*tabula rasa*) at birth, and that all knowledge is gained by sense experience. Berkeley (Fig. 5.3) considered the frame problem: i.e., the fact that as we move around a particular object its perspective changes, and yet we as humans still manage to recognise it as the same object. This problem is relevant in machine vision in AI.

Fig. 5.3 George Berkely Bishop of Cloyne

[14] I think, therefore I am.

[15] Berkeley was an Irish philosopher (not British). He was born in Dysart castle in Kilkenny, Ireland; educated at Trinity College, Dublin; and served as bishop of Cloyne in Co. Cork. He planned to establish a seminary in Bermuda for the sons of colonists in America, but the project failed due to lack of funding from the British government. Berkeley University in San Francisco is named after him.

Berkeley appealed to ideas like the soul to unify the various appearances of the chair as the observer moves around the room, and to God to keep in existence things that were not being perceived.[16] Berkely argued that the ideas in a man's mind have no existence outside his mind [Ber:99], and this philosophical position is known as Idealism. David Hume formulated the standard empiricist philosophical position in his book "An Enquiry concerning Human Understanding" [Hum:06], and this was published in 1748.

Hume (Fig. 5.4) distinguished between ideas and impressions. The term "*impression*" means the lively perceptions that a human directly experiences when hearing, seeing, experiencing anger, and so on. Ideas are less lively perceptions and are experienced when reflecting on sensations such as hearing, seeing and so on. Hume argued that every idea is copied from some impression, and that it is not possible for an idea to arise without its corresponding impression. Hume identified three relations between ideas: resemblance (i.e., similarity), contiguity (experienced together or otherwise) in time or space, and cause or effect.

The empiricist position formulated by Hume argued that all objects of human knowledge may be divided into two kinds (Table 5.1):

Table 5.1 Humes theory of empiricism

Kind of knowledge	Description
Relations of Ideas	These include the truths of propositions in Geometry, Algebra and Arithmetic. They are demonstrated by abstract reasoning through the operation of the human mind. These truths are independent of the outside world, and their denial leads to a contradiction.
Matter of Fact	The evidence for the truth of this kind of knowledge is not as conclusive as knowledge gained through relations of ideas. These truths may not be demonstrated (logically) to be true, and the denial of these propositions does not lead to a contradiction. The evidence for these truths is generally based on experience, and cause and effect. There is a chain of reasons for believing that a particular proposition is true, and these reasons are based on experience.

Hume argued that all knowledge consists of either matters of fact or propositions that may be demonstrated (such as geometry) via deduction reasoning in the operations of the mind.

---

[16] Berkeley's theory of Ontology is that for an entity to exist it must be perceived: i.e., "*Esse est percipi*". He argues that "It is an opinion strangely prevailing amongst men, that houses, mountains, rivers, and in a world all sensible objects have an existence natural or real, distinct from being perceived".
This led to a famous Limerick that poked fun at Berkeley's theory. "There once was a man who said God; Must think it exceedingly odd; To find that this tree, continues to be; When there is no one around in the Quad".
The reply to this Limerick was appropriately: "Dear sir, your astonishments odd; I am always around in the Quad; And that's why this tree will continue to be; Since observed by, yours faithfully, God".

Fig. 5.4 David Hume

The denial or a matter of fact does not involve a contradiction, and matter of fact propositions may be demonstrated by arguments based on experience or cause and effect. They are based entirely based on experience, and experience provides insight into the nature and bounds of cause and effect, and enables the existence of one object to be inferred from that of another.

Hume [Hum:06] argued that any subject[17] proclaiming knowledge that does not adhere to these empiricist principles should be committed to the flames[18] as such knowledge contains nothing but sophistry and illusion.

Kant's Critique of Pure Reason [Kan:03] was published in 1781 and is a response to Hume's theory of empiricism. Kant argued that there is a third force in human knowledge that provides concepts that can't be inferred from experience. Such concepts include the laws of logic (e.g., modus ponens), causality, and so on, and Kant argued that the third force was the manner in which the human mind structures its experiences. These structures (or systematic rules by which the mind structures its experience) are called categories, and include entities such as modus ponens.

---

[17] Hume argues that these principles apply to subjects such as Theology as its foundations are in faith and divine revelation which are neither matters of fact nor relations of ideas.

[18] "When we run over libraries, persuaded of these principles, what havoc must we make? If we take in our hand any volume; of divinity or school metaphysics, for instance; let us ask, *Does it contain any abstract reasoning concerning quantity or number?* No. *Does it contain any experimental reasoning concerning matter of fact and existence?* No. Commit it then to the flames: for it can contain nothing but sophistry and illusion".

The continental school of philosophy included thinkers such as Heidegger, and Merleau-Ponty who argued that the world and the human body are mutually intertwined. Heidegger emphasized that existence can only be considered with respect to a changing world. Merleau-Ponty emphasized embodiment: i.e., that human cognition relies heavily on the body, and he identified the concept of a body-subject that actively participates both as the perceiver of knowledge and as an object of perception.

The analytic school of philosophy was an attempt in the early twentieth century to clean up philosophy. The school believed that many of the philosophical problems were due to inadequate scientific knowledge and an abuse of language. The school was founded in Vienna after the first-world war, and it was known as the Vienna circle. They developed what became known as logical empiricism and this employed the verification principle. This principle stated that a proposition could have meaning if and only if it could be verified. This principle was later shown to be flawed and logical empiricism failed.

Wittgenstein was living in Vienna at the time, and although he was not a member of the circle, he had done important work in his Tractatus Logico Philosophicus [Wit:22] to produce a full deconstruction of language into atomic propositions. These atomic propositions refer to simple objects in the world, which could then be combined to form more complex objects.

Philosophy has been studied for over two millennia but to date very little progress has been made in solving the fundamental philosophical questions. Artificial Intelligence will need to take a practical approach to dealing with the key philosophical questions.

5.5 Cognitive Psychology

Psychology arose out of the field of psychophysics in the late nineteenth century in which various German pioneers attempted to quantify perception and sensation. In fact, one of these pioneers, Fechner, produced a mathematical formulation of the relationship between stimulus and sensation:

$$S = k \log I + c$$

The symbol S refers to the intensity of the sensation, the symbols k and c are constants, and the symbol I refers to the physical intensity of the stimulus. Psychology was defined by William James as the science of mental life.

One of the early behaviouralist psychologist's was Pavlov who showed that it was possible to develop a conditional reflex in a dog. His experiment showed that it was possible to make a dog salivate in response to the ringing of a bell. He did this by ringing a bell each time before meat was provided to the dog, and the dog therefore associated the presentation of meat with the ringing of the bell after a training period.

Skinner developed the concept of conditioning further using rewards to reinforce desired behaviour, and punishment to discourage undesired behaviour. The reinforcement of desired behaviour helps to motivate the individual to behave in a desired way, with punishment used to deter the individual from performing undesired behaviour. The behavioural theory of psychology explains many behavioural aspects of the world, although it does not really explain aspects of complex learning such as language development. Further, behaviouralism does not take human experience into account, and experience plays an important role in explaining human behaviour.

Merleau-Ponty[19] considered the problem of what the structure of the human mind must be in order to allow the objects of the external world to exist in our minds in the form that they do. He argued for phenomenology as originally developed by Hegel and Husserl. This involves a focus and exploration of phenomena with the goal of establishing the essential features of experience. Merleau-Ponty developed the concept of phenomenology further by introducing the concept of the body-subject. This concept is distinct from the Cartesian view that the world is just an extension of our own mind, and instead takes the position that the world and the human body are mutually intertwined. That is, in the Cartesian viewpoint it is first necessary for the self to be aware of and to recognize its own existence prior to being aware of and recognizing the existence of anything else. The body-subject concept expresses the embodied feeling of reality that all humans share.

Merleau-Ponty argued that the body plays a key role in the process of human understanding. The body has the ability to perceive the world and it plays a double role in that it is both the subject (i.e., the perceiver) and the object (i.e., the entity being perceived) of experience. Human understanding and perception is dependent on the body's capacity to perceive via the senses, and its ability to interrogate its environment. Merleau-Ponty argued that there is a symbiotic relationship between the perceiver and what is being perceived, and he argues that as our consciousness develops the self imposes richer and richer meanings on objects. He provides a detailed analysis of the flow of information between the body-subject and the world.

Cognitive psychology is a branch of psychology that is concerned with learning, language, memory and internal mental processes. Its roots lie in the child development psychology developed by Piaget, and in the Gesalt psychology developed by Wertheimer. The latter argues that the operations of the mind are holistic, and that the mind contains a self-organising mechanism. The philosophical position known as Holism argues that the sum of the parts is less than the whole subject, and it is the opposite of the philosophical position of logical Atomism[20] developed by Bertrand Russell. Russell (and also Wittgenstein) attempted to identify the atoms of thought: i.e., the elements of thought that cannot be divided into smaller pieces. The philosophical position of logical atomism was that all truths are ultimately dependent on

[19] Merleau-Ponty was a French philosopher who was strongly influenced by the phenomenology of Husserl. He was also closely associated with the French existentialist philosophers such as Jean-Paul Sartre and Simone De Beauvoir.

[20] Atomism actually goes back to the work of the ancient Greeks and was originally developed by Democritus and his teacher Leucippus in the fifth century BC. Atomism was rejected by Plato in the dialogue the Timaeus.

a layer of atomic facts. It had an associated methodology whereby by a process of analysis it attempted to construct more complex notions in terms of simpler ones.

Cognitive psychology is concerned with how people understand, diagnose, and solve problems, and the mental processes that take place during a stimulus and its corresponding response. It argues that solutions to problems take the form of rules, heuristics and sudden insight. It was developed in the late 1950s, and it views the mind as having a certain conceptual structure. Cognition is considered as the processes by which the sensory input is transformed and used. The dominant paradigm in the field has been the information processing model, and this viewpoint considers the mental processes of thinking and reasoning as being equivalent to the software running on the computer: i.e., the brain. The information processing model has associated theories of input, representation of knowledge, processing of knowledge and output.

Cognitive psychology has been applied to artificial intelligence from the 1960s, and it has played an important role in the integration of various approaches to the mind and mental processes. Some of the research areas in cognitive psychology include:

- Perception
- Concept Formation
- Memory
- Knowledge Representation
- Learning
- Language
- Grammar and Linguistics
- Thinking
- Logic and Problem Solving

The success of these research areas in cognitive psychology will strongly influence developments in Artificial Intelligence. For a machine to behave with intelligence it must be able to perceive objects of the physical world and form concepts about the world. It must be able to remember knowledge that it has already been provided with, as well as possessing an understanding of temporal events. There must be a way to represent knowledge in a machine that allows the knowledge to be easily retrieved and used in analysis and decision making. For a machine to exhibit intelligence it must possess an understanding of language. That is, it must be capable of understanding written language as well as possessing the ability to communicate in written or audio form. A thinking machine must be capable of thought, analysis and problem solving.

5.6 Linguistics

Linguistics is the theoretical or applied scientific study of language. Theoretical linguistics is concerned with the study of syntax and semantics, whereas applied linguistics is concerned with putting linguistic theories into practice.

The study of syntax includes the study of the rules of grammar, and the grammar determines the way by which valid sentences and phrases are produced in the language. It also includes the study of morphology and phonetics. Morphology is concerned with the formation and alteration of words, and phonetics is concerned with the study of sounds, and how sounds are produced and perceived as speech (or non-speech).

Computational Linguistics is the study of the design and analysis of natural language processing systems. It is an interdisciplinary field and includes linguists, computer scientists, experts in artificial intelligence, cognitive psychologists, and mathematicians. Human language is highly complex, and this is reflected in the sophistication of human intelligence.

Computational linguistics originated in the United States in the 1950s in an effort to design software to translate texts written in foreign languages into English. The objective at that time was to develop an automated mechanism by which Russian language texts could be translated directly into English without human intervention. It was believed that given the speed and power of the emerging computer technology that it was only a matter of time before it would be possible by design software to analyse Russian texts and to automatically provide a corresponding English translation.

However, the initial results of automated machine translation of human languages were not very successful, and it was realised that the automated processing of human languages was considerably more complex than initially believed. This led to the birth of a new field called computational linguistics, and the objective of this field is to investigate and to develop algorithms and software for processing natural languages. It is a sub-field of artificial intelligence, and deals with the comprehension and production of natural languages.

The task of translating one language into another requires an understanding of the grammar of both languages. This includes an understanding of the syntax, the morphology, semantics and pragmatics of the language. For Artificial Intelligence to become a reality it will need to make major breakthroughs in computational linguistics.

5.7 Cybernetics

The interdisciplinary field of cybernetics[21] began in the late 1940s when concepts such as information, feedback, and regulation were generalized from engineering to other systems. These include systems such as living organisms, machines, robots and language. The term "*cybernetics*" was coined by Norma Wiener, and it taken from the Greek word "κυβερνήτης" (meaning steersman or governor).

---

[21] Cybernetics was defined by Couffignal (one of its pioneers) as the art of ensuring the efficacy of action.

The name is appropriate as a steersman has a set of goals in traveling to a particular destination, and will need to respond to different conditions on the journey. Similarly, the field of cybernetics is concerned with an interaction of goals, predictions, actions, feedback, and response in systems of all kinds. It is the study of communications and control and feedback in living organisms and machines to ensure efficient action. Practitioners of cybernetics use models of organisations, feedback and goals to understand the capacity and limits of any system.

Cybernetics is concerned with approaches to acquire knowledge through control and feedback, and it is a branch of applied epistemology. The principles of cybernetics are similar to the way that humans acquire knowledge (e.g., learning is achieved through a continuous process of transformation of behaviour through feedback from parents and peers, rather than an explicit encoding of knowledge). The systems studied include observed systems and observing systems (systems that explicitly incorporate the observer into the system).

AI is based on the assumption that knowledge is something that can be stored inside a machine, and that the application of stored knowledge to the real world in this way constitutes intelligence. External objects are mapped to internal states on the machine, and machine intelligence is manifested by the manipulation of the internal states. This approach has been reasonably successful with rule-based expert systems. However, the progress of AI in creating intelligent machines has not been very successful to date, and an alternative approach based on cybernetics may become more important in the future.

Cybernetics views information (or intelligence) as an attribute of an interaction, rather than something that is stored in a computer. The concepts of cybernetics have become part of other disciplines, and it may have an important role to play in the Artificial Intelligence field.

5.8 Logic and AI

Logic is a key discipline in the AI field, and the goal is to use mathematical logic to formalise knowledge and reasoning to enable AI problems to be solved. John McCarthy and others have argued that mathematical logic has a key role to play in the formalisation of common-sense knowledge and reasoning. Common-sense knowledge includes basic facts about actions and their effects, facts about beliefs and desires, and facts about knowledge and how it is obtained. He argues that common-sense reasoning is required for solving problems in the real world, and that therefore it is reasonable to apply common-sense knowledge to the world of artificial intelligence. His approach is to use mathematical logic to formalise common-sense knowledge to allow common-sense problems to be solved by logical reasoning.

This requires sufficient understanding of the common-sense world to enable facts to be formalised. One problem is that often the reasoner doesn't know which facts are relevant in solving a particular problem. Knowledge that was thought to be irrelevant to the problem to be solved may be essential. A computer may have millions of facts stored in its memory, and the problem is how can it determine the relevant facts from its memory to serve as premises for logical deduction.

McCarthy philosophy of producing programs with common sense was discussed in his influential 1959 paper [Mc:59]. This paper discusses various common-sense problems such as getting home from the airport. Other examples of common-sense problems are diagnosis, spatial reasoning, and understanding narratives that include temporal events.

Computers need a precise language to store facts about the world and to reason about them. It needs a rigorous definition of the reasoning that is valid as this will determine how the deduction of a new formula from a set of formulae is done. Mathematical logic is the standard approach to express premises, and it includes rules of inferences that are used to deduce valid conclusions from a set of premises.

The propositional and predicate calculii were discussed earlier in Chapter . Propositional calculus associates a truth-value with each proposition, and includes logical connectives to produce formulae such as $A \Rightarrow B$, $A \wedge B$, and $A \vee B$. The truth values of the propositions are normally the binary values of *true* and *false*. There are other logics, such as 3-valued logic or fuzzy logics that allow more than two truth-values for a proposition. Predicate logic is more expressive than propositional logic, and it can formalise the syllogism "All Greeks are mortal; Socrates is a Greek; Therefore, Socrates is mortal". The predicate calculus consists of:

- Axioms
- Rules for defining well-formed formulae
- Inference rules for deriving theorems from premises

A formula in predicate calculus is built up from the basic symbols of the language. These include variables, predicate symbols such as equality; function symbols, constants logical symbols such as: $\exists$, $\wedge$, $\vee$, $\neg$, and punctuation symbols such as brackets and commas. The formulae of predicate calculus are built from terms, where a *term* is defined recursively as a variable or individual constant or as some function containing terms as arguments. A formula may be an atomic formula, or built from other formulae via the logical symbols.

There are several rules of inference associated with predicate calculus, and the most important of these are modus ponens and generalisation. The rule of modus ponens states that given two formulae p, and $p \Rightarrow q$, then we may deduce q. The rule of generalisation states that given a formula p that we may deduce $\forall(x)p$.

McCarthy's approach to programs with common sense has been criticized by Bar-Hillel and others on several grounds:

- How can a computer know from the set of facts that it knows which facts are appropriate to apply in deduction in a given situation.
- Many philosophers believe that common sense is fairly elusive.

However, while logic is unlikely to prove the main solution to the problems in AI, it is likely to form part of the solution.

5.9 Computability, Incompleteness and Decidability

An algorithm (or procedure) is a finite set of unambiguous instructions to perform a specific task. The term "algorithm" is named after the Persian mathematician Al-Khwarizmi. The concept of an algorithm was defined formally by Church in 1936 with the lambda calculus, and independently by Turing with the Turing machines. Turing defined computability in terms of the mechanical procedure of a Turing machine, whereas Church defined computability in terms of the lambda calculus. Later it was shown that computability by Turing machines and lambda calculus are equivalent.

Formalism was proposed by Hilbert as a foundation for mathematics in the early twentieth century. A formal system consists of a formal language, a set of axioms and rules of inference. Hilbert's programme was concerned with the formalization of mathematics (i.e., the axiomatization of mathematics) together with a proof that the axiomatization was consistent. The specific objectives of Hilbert's programme were to:

- Develop a formal system where the truth or falisty of any mathematical statement may be determined.
- A proof that the system is consistent (i.e., that no contradictions may be derived).

A proof in a formal system consists of a sequence of formulate, where each formula is either an axiom or derived from one or more preceding formulae in the sequence by one of the rules of inference. Hilbert believed that every mathematical problem could be solved and therefore expected that the formal system of mathematics would be completes (that is, all truths could be proved within the system) and decidable: i.e., that the truth or falisty of any mathematical proposition could be determined by an algorithm.

Russell and Whitehead published Principia Mathematica in 1910, and this three-volume work on the foundations of mathematics attempted to derive all mathematical truths in arithmetic from a well-defined set of axioms and rules of inference. The questions remained whether the Principia was *complete* and *consistent.* That is, is it possible to derive all the truths of arithmetic in the system and is it possible to derive a contradiction from the Principia's axioms?

Gödel's second incompleteness theorem [Goe:31] showed that first order arithmetic is incomplete, and that the consistency of first-order arithmetic cannot be proved within the system. Therefore, if first-order arithmetic cannot prove its own consistency, then it cannot prove the consistency of any system that contains first-order arithmetic.

Hilbert also believed that formalism would be decidable: there would be a mechanical procedure (or algorithm) to determine whether a particular statement was true or false. However, Church and Turing independently showed this to be impossible in 1936. The only way to determine whether a statement is true or false is to try to solve it.

5.10 Robots

The first use of the term "robot" was by the Czech playwright Karel Capek in his play "*Rossum's Universal Robots*" performed in Prague in 1921. The word "robot" is from the Czech word for forced labour. The play was discussed earlier in Section 5.5, and the theme explored is whether it is ethical to exploit artificial workers in a factory, and what response the robots should make to their exploitation. Capek's robots were not mechanical or metal in nature and were instead created through chemical means. Capek, in fact, rejected the idea that machines created from metal could think of feel.

The science fiction writer Asimov wrote several stories about robots in the 1940s including the story of a robotherapist.[22] He predicted the rise of a major robot industry, and he also introduced a set of rules (or laws) which stipulated the good behaviors that robots are expected to observe. These are known as the three Laws of Robotics and a fourth law was later added by Asimov (Table 5.2).

The term "robot" is defined by the Robot Institute of America as:

Definition 5.1 (Robots) A re-programmable, multifunctional manipulator designed to move material, parts, tools, or specialized devices through various programmed motions for the performance of a variety of tasks.

Joseph Engelberger and George Devol are considered the fathers of robotics. Engelberger set up the first manufacturing company "Unimation" to make robots, and Devol wrote the necessary patents. Their first robot was called the "Unimate". These robots were very successful and reliable, and saved their customer (General Motors) money by replacing staff with machines. The robot industry continues to play a major role in the automobile sector.

Robots have been very effective at doing clearly defined repetitive tasks, and there are many sophisticated robots in the workplace today. These robots are industrial manipulators and are essentially computer controlled "arms and hands". However, fully functioning androids are many years away.

Table 5.2 Laws of Robotics

Law	Description
Law Zero	A robot may not injure humanity, or, through inaction, allow humanity to come to harm.
Law One	A robot may not injure a human being, or, through inaction, allow a human being to come to harm, unless this would violate a higher order law.
Law Two	A robot must obey orders given it by human beings, except where such orders would conflict with a higher order law.
Law Three	A robot must protect its own existence as long as such protection does not conflict with a higher order law.

[22] The first AI therapist was the ELIZA program produced by Weizenbaum in the mid-1960s.

Robots can also improve the quality of life for workers as they can free human workers from performing dangerous or repetitive jobs. Further, it leads to work for robot technicians and engineers. Robots provide consistently high-quality products and can work tirelessly 24 hours a day. This helps to reduce the costs of manufactured goods thereby benefiting consumers. They do not require annual leave but will, of course, from time to time require servicing by engineers or technicians. However, there are impacts on workers whose jobs are displaced by robots.

5.11 Neural Networks

The term "neural network" (artificial or biological) refers to an interconnected group of processing elements called nodes or neurons. These neurons cooperate and work together to produce an output function. Neural networks may be artificial or biological. A biological network is part of the human brain, whereas an artificial neural network is designed to mimic some properties of a biological neural network. The processing of information by a neural network is done in parallel rather than in series. A unique property of a neural network is fault tolerance: i.e., it can still perform (within certain tolerance levels) its overall function even if some of its neurons are not functioning. There are trainable neural network systems that can learn to solve complex problems from a set of examples. These systems may also use the acquired knowledge to generalise and solve unforeseen problems.

A biological neural network is composed of billions of neurons (or nerve cells). A single neuron may be physically connected to thousands of other neurons, and the total number of neurons and connections in a network may be extremely large. The human brain consists of many billions of neurons, and these are organized into a complex intercommunicating network. The connections are formed through axons[23] to dendrites,[24] and the neurons can pass electrical signals to each other. These connections are not just the binary digital signals of *on* or *off*, and instead the connections have varying strength which allows the influence of a given neuron on one of its neighbors to vary from very weak to very strong.

That is, each connection has an individual weight (or number) associated with it that indicates its strength. Each neuron sends its output value to all other neurons to which it has an outgoing connection. The output of one neuron can influence the activations of other neurons causing them to fire. The neuron receiving the connections calculates its activation by taking a weighted sum of the input signals. The output is determined by the activation function based on this activation. Networks learn by changing the weights of the connections. Many aspects of brain function, especially the learning process, are closely associated with the adjustment of these connection strengths. Brain activity is represented by particular patterns of firing

[23] These are essentially the transmission lines of the nervous system. They are microscopic in diameter and conduct electrical impulses.

[24] Dendrites are in effect the cell body and extend like the branches of a tree. The origin of the word dendrite is from the Greek word for tree.

activity amongst the network of neurons. This simultaneous cooperative behavior of a huge number of simple processing units is at the heart of the computational power of the human brain

Artificial neural networks aims to simulate various properties of biological neural networks. These are computers whose architecture is modeled on the brain. They consist of many hundreds of simple processing units which are wired together in a complex communication network. Each unit or node is a simplified model of a real neuron which fires[25] if it receives a sufficiently strong input signal from the other nodes to which it is connected. The strength of these connections may be varied in order for the network to perform different tasks corresponding to different patterns of node firing activity. The objective is to solve a particular problem, and artificial neural networks have been successfully applied to speech recognition problems, image analysis, and so on. Many of the existing artificial neural networks are based on statistical estimation and control theory. Artificial neural networks have also been applied to the cognitive modelling field.

There are similarities between the human brain and a very powerful computer with advanced parallel processing. Artificial neural networks have provided simplified models of the neural processing that takes place in the brain. The challenge for the field is to determine what properties individual neural elements should have to produce something useful representing intelligence.

Neural networks are quite distinct from the traditional von Neumann architecture[26] described in Chapter . The latter is based on the sequential execution of machine instructions. The origins of neural networks lie in the attempts to model information processing in biological systems. This relies more on parallel processing as well as on implicit instructions based on pattern recognition from sense perceptions of the external world. The nodes in an artificial neural network are composed of many simple processing units which are connected into a network. Their computational power depends on working together (parallel processing) on any task, and there is no central processing unit following a logical sequence of rules. Computation is related to the dynamic process of node firings. This structure is much closer to the operation of the human brain, and leads to a computer that may be applied to a number of complex tasks.

5.12 Expert Systems

An expert system is a computer system that contains the subject-specific knowledge of one or more human experts. These systems allow knowledge to be stored and intelligently retrieved. Expert Systems arose in Artificial Intelligence during the

[25] The firing of a neuron means that it sends off a new signal with a particular strength (or weight).

[26] In fact, many computers today still have an underlying von Neumann architecture. However, there have been major advances in speed and size of the hardware since the 1940s. Computers consist of a Central Processing Unit which executes the software (set of rules), and any intelligence in the machine resides in the set of rules supplied by the programmer.

1960s, and several commercial expert systems have been developed since then. An expert system is a program made up of a set of rules (or knowledge) supplied by the subject-matter experts about a specific class of problems. The success of the expert system is largely dependent on the quality of the rules provided by the expert. The expert system also provides a problem solving component that allows analysis of the problem to take place, as well as recommending an appropriate course of action to solve the problem. That is, the expert system employs a reasoning capability to draw conclusions from facts that it knows, and it also recommends an appropriate course of action to the user.

Expert Systems have been a major success story in the AI field. They have been employed in various areas such as for diagnostics in the medical field, equipment repair, and investment analysis. There is a clear separation between the collection of knowledge and the problem solving strategy of the system. Expert systems consist of the following components (Table 5.3):

Table 5.3 Expert systems

Component	Description
Knowledge Base	The knowledge base is composed by experts in a particular field, and the expert system is only as good as the knowledge provided by the experts.
Communication Components	This component (typically a user interface) allows the user to communicate with the expert system. It allows the user to pose questions to the system and to receive answers (or vice versa).
Context Sensitive Help	This is usually provided on all screens and the help provided is appropriate to where the user is in the system.
Problem Solving Component	The problem solving component analyses the problem and uses its inference engine to deduce the appropriate course of action to resolve the problem.
	The inference engine is software that interacts with the user and the rules in the knowledge base to produce an appropriate solution.

Expertise consists of knowledge about a particular domain and involves understanding problems in the domain as well as skill at solving these problems. Human knowledge of a specialty is of two types: namely public knowledge and private knowledge. The former includes the facts and theories documented in text books and publications on the domain, whereas the latter refers to knowledge that the expert possesses that has not found its way into the public domain. The latter often consists of rules of thumb or heuristics that allow the expert to make an educated guess where required, as well as allowing the expert to deal effectively with incomplete or erroneous data. The challenge for the AI field is to allow the representation of both public and private knowledge to enable the expert system to draw valid inferences.

The inference engine is made up of many inference rules that are used by the engine to draw conclusions. Rules may be added or deleted without affecting other rules, and this reflects the normal updating of human knowledge. That is, out-dated facts are, in effect, deleted and are no longer used in reasoning, while new knowledge is applied in drawing conclusions. The inference rules use reasoning which is closer to human reasoning. There are two main types of reasoning with inference rules and these are backward chaining and forward chaining. Forward chaining starts with the data available, and uses the inference rules to draw intermediate conclusions until a desired goal is reached. Backward chaining starts with a set of goals and works backwards to determine if one of the goals can be met with the data that is available.

The expert system makes its expertise available to decision makers who need answers quickly. This is extremely useful as often there is a shortage of experts, and the availability of an expert computer with in-depth knowledge of specific subjects is therefore very attractive. Expert systems may also assist managers with long-term planning. There are many small expert systems available that are quite effective in a narrow domain. The long-term goal is to develop expert systems with a broader range of knowledge. Expert systems have enhanced productivity in business and engineering, and there are several commercial software packages available to assist.

Some of the well-known expert systems that have been developed include Mycin, Colossus and Dendral. Mycin was developed at Stanford University in the 1970s, and it was developed from the original Dendral system. It was written in LISP and it was designed to diagnose infectious blood diseases, and to recommend appropriate antibiotics and dosage amounts corresponding to the patient's body weight. Mycin had a knowledge base of approximately 500 rules and had a fairly simple inference engine. Its approach was to query the physician running the program with a long list of yes/no questions. Its output consisted of various possible bacteria that could correspond to the blood disease, along with an associated probability that indicated the confidence in the diagnosis, and it also included the rationale for the diagnosis and a course of treatment appropriate to the diagnosis.

Mycin's performance was reasonable as it had a correct diagnosis rate of 65%. This was better than the diagnosis of most physicians who did not specialise in bacterial infections. However, its diagnosis rate was less than that of experts in bacterial infections who had a success rate of 80%. Mycin was never actually used in practice due to legal and ethical reasons in the use of expert systems in medicine. For example, if the machine makes the wrong diagnosis who is to be held responsible?

Colossus was an expert system used by several major Australian insurance companies. It helps insurance adjusters to assess personal injury claims, and helps to improve consistency in the claims process. It aims to make claims fair and objective by guiding the adjuster through an objective evaluation of medical treatment options, the degree of pain and suffering of the claimant, and the extent that there is permanent impairment to the claimant, as well as the impact of the injury on the claimant's lifestyle. It was developed by Computer Sciences Corporation (CSC).

Dendral (Dendritic Algorithm) was developed at Stanford University in the mid-1960s, and it was the first use of Artificial Intelligence in medical research. Its objectives were to assist the organic chemist with the problem of identifying unknown organic compounds and molecules by computerized spectrometry. This involved the analysis of information from mass spectrometry graphs and knowledge of chemistry. Dendral automated the decision-making and problem-solving process used by organic chemists to identify complex unknown organic molecules. It was written in LISP and it showed how an expert system could employ rules, heuristics and judgment to guide scientists in their work.

5.13 Review Questions

1. Discuss Descartes and his Rationalist philosophy and his relevance to Artificial Intelligence.
2. Discuss the Turing Test and its relevance on Strong AI. Discuss Searle's Chinese Room rebuttal arguments. What are you own views on Strong AI?
3. Discuss the philosophical problems underlying Artificial Intelligence.
4. Discuss the applicability of Logic to Artificial Intelligence.
5. Discuss Neural Networks and their applicability to Artificial Intelligence.
6. Discuss Expert Systems.

5.14 Summary

The origin of the term "Artificial Intelligence" was a proposal for the Dartmouth Summer Research Project on Artificial Intelligence in the mid-1950s. The ultimate goal of Artificial Intelligence is to create a thinking machine that is intelligent, has consciousness, has the ability to learn, has free will, and is ethical. It is a multi-disciplinary field, and its branches include:

- Computing
- Logic and Philosophy
- Psychology
- Linguistics
- Machine Vision
- Computability
- Epistemology and Knowledge representation

Turing believed that machine intelligence was achievable and he devised the famous "Turing Test" to judge whether a machine was conscious and intelligent. Searle's Chinese Room argument is a rebuttal that aims to demonstrate that a machine will never have the same cognitive qualities as a human, and that even if a machine passes the Turing Test it still lacks intelligence and consciousness.

McCarthy's approach to Artificial Intelligence is to use logic to describe the manner in which intelligent machines or people behave. His philosophy is that common sense knowledge and reasoning can be formalised with logic. That is, human-level intelligence may be achieved with a logic-based system.

Cognitive psychology is concerned with cognition and some of its research areas include perception, memory, learning, thinking, and logic and problem solving. Linguistics is the scientific study of language and includes the study of syntax and semantics.

Artificial neural networks aims to simulate various properties of biological neural networks. These are computers whose architecture is modeled on the brain. They consist of many hundreds of simple processing units which are wired together in a complex communication network. Each unit or node is a simplified model of a real neuron which fires if it receives a sufficiently strong input signal from the other nodes to which it is connected. The strength of these connections may be varied in order for the network to perform different tasks corresponding to different patterns of node firing activity. The objective is to solve a particular problem, and artificial neural networks have been successfully applied to speech recognition problems and image analysis.

An expert system is a computer system that allows knowledge to be stored and intelligently retrieved. It is a program that is made up of a set of rules (or knowledge). These rules are generally supplied by the subject-matter experts about a specific class of problems. The expert system also provides a problem solving component that allows analysis of the problem to take place, as well as recommending an appropiate course of action to solve the problem. Expert Systems have been a major success story in the AI field.

Chapter 6
The Internet Revolution

Key Topics

ARPANET
TCP/IP
The Internet
The World-Wide Web
Dot Com Bubble
E-Software Development

6.1 Introduction

The origins of what has developed to become the internet and world wide web goes back to work done in the early 1940s by Vanevar Bush (Fig. 6.1). Bush was an American scientist who had done work on submarine detection for the United States Navy, He later did work with others did work on an automated network analyser to solve differential equations at MIT. He became director of the office of Scientific Research and Development which was concerned with weapons research and development. This organisation employed several thousand scientists, and was responsible for supervising the development of the atomic bomb.

Bush developed a close win-win relationship between the United States military and universities. He arranged large research funding for universities to carry out research related to the needs of the United States military. This allowed the military to benefit from the early exploitation of research results, and it also led to better facilities and laboratories in universities to carry out research. Bush initially developed close links with universities such as Harvard and Berkeley. This was the foundation for future cooperation between the universities and the United States Department of Defence, and it would lead eventually to the development of ARPANET by DARPA.

Fig. 6.1 Vanevar Bush

Bush's influence on the development of the internet is due to his visionary description of an information management system that he called the "*memex*". The memex (memory extender) is described in his famous essay "As We May Think" which was published in the Atlantic Monthly in 1945 [Bus:45]. Bush envisaged the memex as a device electronically linked to a library and able to display books and films.

Bush's article essentially describes a theoretical proto-hypertext computer system, and this influenced the subsequent development of hypertext systems. The description of a memex is described in [Bus:45]:

> Consider a future device for individual use, which is a sort of mechanized private file and library. It needs a name, and to coin one at random, "memex" will do. A memex is a device in which an individual stores all his books, records, and communications, and which is mechanized so that it may be consulted with exceeding speed and flexibility. It is an enlarged intimate supplement to his memory.
>
> It consists of a desk, and while it can presumably be operated from a distance, it is primarily the piece of furniture at which he works. On the top are slanting translucent screens, on which material can be projected for convenient reading. There is a keyboard, and sets of buttons and levers. Otherwise it looks like an ordinary desk.

Bush predicted that:

> Wholly new forms of encyclopedias will appear, ready made with a mesh of associative trails running through them, ready to be dropped into the memex and there amplified..

This description motivated Ted Nelson and Douglas Engelbart to independently formulate the various ideas that would became hypertext. Tim Berners-Lee would later use hypertext as part of the development of the world-wide web.

6.2 The ARPANET

There were approximately 10,000 computers in the world in the 1960s. These computers were very expensive ($100K–$200K) and had very primitive processing power. They contained only a few thousand words of magnetic memory, and programming and debugging of these computers was difficult. Further, communication between computers was virtually non-existent.

However, several computer scientists had dreams of world wide networks of computers, where every computer around the globe is interconnected to all of the other computers in the world. For example, Licklider[1] wrote memos in the early 1960s on his concept of an intergalactic network. This concept envisaged that everyone around the globe would be interconnected and able to access programs and data at any site from anywhere.

The United States Department of Defence founded the Advance Research Projects Agency (ARPA) in the late 1950s. ARPA embraced high-risk, high-return research and laid the foundation for what became ARPANET and later the internet. Licklider became the first head of the computer research program at ARPA, which was called the Information Processing Techniques Office (IPTO). He developed close links with MIT, UCLA, and BBN Technologies[2] and started work on his vision. Various groups, including National Physical Laboratory (NPL), the RAND Corporation and MIT, commenced work on packet switching networks. The concept of packet switching was invented by Donald Davies[3] at the NPL in 1965.

The early computers had different standards for representing data and this meant that the data standard of each computer would need to be known for effective communication to take place. There was a need to establish a standard for data representation, and a United States government committee developed the ANSII (American

---

[1] Licklider was an early pioneer of AI and he also formulated the idea of a global computer network. He wrote his influential paper "*Man–Computer Symbiosis*" in 1960, and this paper outlined the need for simple interaction between users and computers.

[2] BBN Technologies (originally Bolt Beranek and Newman) is a research and development high technology company. It is especially famous for its work in the development of packet switching for the ARPANET and the Internet. It also did defense work for DARPA. BBN played an important part in the implementation and operation of ARPANET. The "@" sign used in an email address was a BBN innovation.

[3] Packet switching is a fast message communication system between computers. Long messages are split into packets which are then sent separately so as to minimise the risk of congestion. Davies also worked on the ACE computer (one of the earliest stored program computers) that was developed at the NPL in the UK in the late 1940s.

Standard Code for Information Interchange) in 1963. This became the first universal standard for data for computers, and it allowed machines from different manufacturers to exchange data. The standard allowed 7-bit binary strings to stand for a letter in the English alphabet, an Arabic numeral or a punctuation symbols. The use of 7 bits allowed 128 distinct characters to be represented. The development of the IBM-360 mainframe standardised the use of 8-bits for a word, and 12-bit or 36-bit words became obsolete.

The first wide-area network connection was created in 1965. It involved the connection of a computer at MIT to a computer in Santa Monica via a dedicated telephone line. This result showed that telephone lines could be used for the transfer of data although they were expensive in their use of bandwidth. The need to build a network of computers became apparent to ARPA in the mid-1960s, and this led to work commencing on the ARPANET project in 1966. The plan was to implement a packet switched network based on the theoretical work done on packet switching done at NPL and MIT. The goal was to have a network speed of 56 Kbps. ARPANET was to become the world's first packet switched network.

BBN Technologies was awarded the contract to implement the network. Two nodes were planned for the network initially and the goal was to eventually have nineteen nodes. The first two nodes were based at UCLA and SRI. The network management was performed by interconnected "Interface Message Processors" (IMPs) in front of the major computers. Each site had a team to produce the software to allow its computers and the IMP to communicate. The IMPs eventually evolved to become the network routers that are used today. The team at UCLA called itself the Network Working Group, and saw its role as developing the Network Control Protocol (NCP). This was essentially a rule book that specified how the computers on the network should communicate.

The first host to host connection was made between a computer in UCLA and a computer in SRI in late 1969. Several other nodes were added to the network until it reached its target of 19 nodes in 1971. The Network Working Group developed the telnet protocol and file transfer protocol (FTP) in 1971. The telnet program allowed the user of one computer to remotely log in to the computer of another computer. The file transfer protocol allows the user of one computer to send or receive files from another computer. A public demonstration of ARPANET was made in 1972 and it was a huge success. One of the earliest demos was that of Weizenbaum's ELIZA program. This is a famous AI program that allows a user to conduct a typed conversation with an artificially intelligent machine (psychiatrist) at MIT. The viability of packet switching as a standard for network communication had been clearly demonstrated.

Ray Tomlinson of BBN Technologies developed a program that allowed electronic mail to be sent over the ARPANET. Tomlinson developed the "user@host" convention, and this was eventually to become the standard for electronic mail in the late 1980s.

By the early 1970s over thirty institutions were connected to the ARPANET. These included consulting organisations such as BBN, Xerox, and the MITRE Corporation, and government organisations such as NASA. Bob Metacalfe developed a wire-based local area network (LAN) at Xerox that would eventually become Ethernet in the mid-1970s.

6.3 TCP/IP

ARPA became DARPA (Defence Advance Research Projects Agency) in 1973, and it commenced work on a project connecting seven computers on four islands using a radio based network, as well as a project to establish a satellite connection between a site in Norway and in the United Kingdom. This led to a need for the interconnection of the ARPANET with other networks. The key problems were to investigate ways of achieving convergence between ARPANET, radio-based networks, and the satellite networks, as these all had different interfaces, packet sizes, and transmission rates. Therefore, there was a need for a network to network connection protocol, and its development would prove to be an important step towards developing the internet.

An international network working group (INWG) was formed in 1973. The concept of the transmission control protocol (TCP) was developed at DARPA by Bob Kahn and Vint Cerf, and they presented their ideas at an INWG meeting at the University of Sussex in England in 1974 [KaC:74]. TCP allowed cross network connections, and it began to replace the original NCP protocol used in ARPANET. However, it would take some time for the existing and new networks to adopt the TCP protocol.

TCP is a set of network standards that specify the details of how computers communicate, as well as the standards for interconnecting networks and computers. It was designed to be flexible and provides a transmission standard that deals with physical differences in host computers, routers, and networks. TCP is designed to transfer data over networks which support different packet sizes and which may sometimes lose packets. It allows the inter-networking of very different networks which then act as one network.

The new protocol standards were known as the transport control protocol (TCP) and the internet protocol (IP). TCP details how information is broken into packets and re-assembled on delivery, whereas IP is focused on sending the packet across the network. These standards allow users to send electronic mail or to transfer files electronically, without needing to concern themselves with the physical differences in the networks. TCP/IP is a family or suite of protocols, and it consists of four layers (Table 6.1):

The internet protocol (IP) is a connectionless protocol that is responsible for addressing and routing packets. It is responsible for breaking up and assembling packets, with large packets broken down into smaller packets when they are travelling through a network that supports smaller packets. A connectionless protocol means that a session is not established before data is exchanged. Packet delivery with IP is not guaranteed as packets may be lost or delivered out of sequence. An acknowledgement is not sent when data is received, and the sender or receiver is not

Table 6.1 TCP layers

Layer	Description
Network Interface Layer	This layer is responsible for formatting packets, and placing them on to the underlying network. It is equivalent to the physical and data link layers on OSI Model.
Internet Layer	This layer is responsible for network addressing. It includes the internet protocol, the address resolution protocol, and so on. It is equivalent to the network layer on the OSI Model.
Transport Layer	This layer is concerned with data transport, and is implemented by TCP and the user datagram protocol (UDP). It is equivalent to the transport and session layers in the OSI Model.
Application Layer	This layer is responsible for liaising between user applications and the transport layer. The applications include the file transfer protocol (FTP), telnet, domain naming system (DNS), and simple mail transfer program (SMTP). It is equivalent to the application and presentation layers on the OSI Model.

informed when a packet is lost or delivered out of sequence. A packet is forwarded by the router only if the router knows a route to the destination. Otherwise, it is dropped. Packets are dropped if their checksum is invalid or if their time to live is zero. The acknowledgement of packets is the responsibility of the TCP protocol. The ARPANET employed the TCP/IP protocols as a standard from 1983.

The late 1970s saw the development of newsgroups that aimed to provide information about a particular subject. A newsgroup started with a name that is appropriate with respect to the content that it is providing. Newsgroups were implemented via USENET, and were an early example of client-server architecture. A user dials in to a server with a request to forward a certain newsgroup postings; the server then "serves" the request.

6.4 Birth of the Internet

The origins of the internet can be traced to the United States government support of the ARPANET project. Computers in several United States universities were linked via packet switching, and this allowed messages to be sent between the universities that were part of the network.

The use of ARPANET was limited initially to academia and to the United States military, and in the early years there was little interest from industrial companies. However, by the mid-1980s there were over 2,000 hosts on the TCP/IP enabled network, and the ARPANET was becoming more heavily used and congested. It was decided to shut down the network by the late-1980s, and the National Science

Foundation in the United States commenced work on the NSFNET. This work commenced in the mid-1980s, and the network consisted of multiple regional networks connected to a major backbone. The original links in NSFNET were 56 Kbps but these were later updated to the faster T1 (1.544 Mbps) links. The NSFNET T1 backbone initially connected 13 sites, but this increased due to a growing interest from academic and industrial sites in the United States and from around the world. The NSF began to realize from the mid-1980s onwards that the internet had significant commercial potential.

The NSFNET backbone was upgraded with T1 links in 1988 and the internet began to become more international. Sites in Canada and several European countries were connected to the internet. DARPA formed the Computer Emergency Response Team (CERT) to deal with any emergency incidents arising during the operation of the network.

Advanced Network Services (ANS) was founded in 1991. This was an independent not-for-profit company, and it installed a new network that replaced the NSFNET T1 network. The ANSNET backbone operated over T3 (45 Mbps) links, and it different from previous networks such as ARPANET and NSFNET in that it was owned and operated by a private company rather than the United States government. The NSF decided to focus on research aspects of networks rather than the operational side. The ANSNET network was a move away from a core network such as NSFET, to a distributive network architecture operated by commercial providers such as Sprint, MCI and BBN. The network was connected by major network exchange points, termed Network Access Points (NAPs). There were over 160,000 hosts connected to the internet by the late 1980s.

The discovery of the world-wide web by Tim Berners-Lee at CERN in 1990 was a revolutionary milestone in computing. It has transformed the way that businesses operate as well as transforming the use of the internet from mainly academic (with some commercial use) to an integral part of peoples' lives. The invention of the world-wide web by Berners-Lee is described in the next section.

6.5 Birth of the World-Wide Web

The world-wide web was invented by Tim Berners-Lee (Fig. 6.2) in 1990 at CERN in Geneva, Switzerland. CERN is a key European and international center for research in the nuclear field, and several thousand physicists and scientists work there. Lee first came to CERN in 1980 for a short contract programming assignment. He came from a strong scientific background as both his parents had been involved in the programming of the Mark I computer at Manchester University in the 1950s. He graduated in physics in the mid-1970s at Oxford University in England.

One of the problems that scientists at CERN faced was that of keeping track of people, computers, documents, databases, etc. This problem was more acute due to the international nature of CERN, as the centre had many visiting scientists from overseas who spent several months there. It also had a large pool of permanent staff.

Fig. 6.2 Tim Berners-Lee Photo courtesy of Wikipedia.

Visiting scientists used to come and go, and in the late 1980s there was no efficient and effective way to share information among scientists.

It was often desirable for a visiting scientist to obtain information or data from the CERN computers. In other cases, the scientist wished to make results of their research available to CERN in an easy manner. Berners-Lee developed a program called "Enquire" to assist with information sharing, and the program also assisted in keeping track of the work of visiting scientists. He returned to CERN in the mid-1980s to work on other projects, and he devoted part of his free time to consider solutions to the information sharing problem. This was eventually to lead to his breakthrough and his invention of the world-wide web in 1990.

Inventors tend to be influenced by existing inventions, and especially inventions that are relevant to their areas of expertise. The internet was a key existing invention, and it allowed world wide communication via electronic email, the transfer of files electronically via FTP, and newsgroups that allowed users to make postings on various topics. Another key invention that was relevant to Berners-Lee was that of hypertext. This was invented by Ted Nelson in the 1960s, and it allow links to be present in text. For example, a document such as a book contains a table of contents, an index, and a bibliography. These are all links to material that is either within the book itself or external to the book. The reader of a book is able to follow the link to obtain the internal or external information.

The other key invention that was relevant to Berners-Lee was that of the mouse. This was invented by Doug Engelbart in the 1960s, and it allowed the cursor to be steered around the screen. The major leap that Berners-Lee made was essentially a marriage of the internet, hypertext and the mouse into what has become the world-wide web. His vision and its subsequent realisation was beneficial both to CERN and the wider world. He described the vision as follows [BL:00]:

> Suppose that all information stored on computers everywhere were linked. Program computer to create a space where everything could be linked to everything.

Berners-Lee essentially created a system to give every "page" on a computer a standard address. This standard address is called the universal resource locator and is better known by its acronym URL. Each page is accessible via the hypertext transfer protocol (HTTP), and the page is formatted with the hypertext markup language (HTML). Each page is visible using a web-browser. The key features of Berners-Lee invention are (Table 6.2):

Table 6.2 Features of world-wide web

Feature	Description
URL	Universal Resource Identifier (later renamed to Universal Resource Locator (URL). This provides a unique address code for each web page. Browsers decode the URL location to access the web page. For example, www.amazon.com uniquely identifies the Amazon.com host web site in the United States.
HTML	Hyper Text Markup Language (HTML) is used for designing the layout of web pages. It allows the formatting of pages containing hypertext links. HTML is standardized and controlled by the World Wide Web Consortium (http://www.w3.org).
HTTP	The Hypertext Transport Protocol (HTTP) allows a new web page to be accessed from the current page.
Browser	A browser is a client program that allows a user to interact with the pages and information on the world-wide web. It uses the HTTP protocol to make requests of web servers throughout the internet on behalf of the browser user. Berners-Lee developed the first web browser called the World Wide Web browser.

Lee invented the well-known terms such as URL, HTML and world-wide web, and these terms are ubiquitous today. He also wrote the first browser program, and this allowed users to access web pages throughout the world. Browsers are used to connect to remote computers over the internet, and to request, retrieve and display the web pages on the local machine. The invention of the world-wide web by Berners-Lee was a revolution in the use of the internet. Users could now surf the web: i.e., hyperlink among the millions of computers in the world and obtain information easily.

The early browsers included Gopher developed at the University of Minnesota, and Mosaic developed at the University of Illinois. These were replaced in later years by Netscape and the objective of its design was to create a graphical-user-interface browser that would be easy to use, and would gain widespread acceptance in the internet community. Initially, the Netscape browser dominated the browser market, and this remained so until Microsoft developed its own browser called Internet Explorer. Microsoft's browser would eventually come to dominate the browser market, after what became known as the browser wars. The eventual dominance of Microsoft internet explorer was controversial, and it was subject to legal investigations in the United States. The development of the graphical browsers led to the commercialisation of the world-wide web.

The world-wide web creates a space in which users can access information easily in any part of the world. This is done using only a web browser and simple web addresses. The user can then click on hyperlinks on web pages to access further relevant information that may be on an entirely different continent. Berners-Lee is now the director of the World Wide Web Consortium, and this MIT based organisation sets the software standards for the Web.

6.6 Applications of the World-Wide Web

Berners-Lee used the analogy of the market economy to describe the commercial potential of the world-wide web. He realized that the world-wide web offered the potential to conduct business in cyberspace without human interaction, rather than the traditional way of buyers and sellers coming together to do business in the market place.

> Anyone can trade with anyone else except that they do not have to go to the market square to do so

The invention of the world-wide web was announced in August 1991, and the growth of the web has been phenomenal since then. The growth has often been exponential, and exponential growth rate curves became a feature of newly formed internet companies, and their business plans. The world-wide web is revolutionary in that:

- No single organization is controlling the web.
- No single computer is controlling the web.
- Millions of computers are interconnected.
- It is an enormous market place of millions (billions) of users.
- The web is not located in one physical location.
- The web is a space and not a physical thing.

The world-wide web has been applied to many areas including:

- Travel Industry (Booking flights, train tickets, and hotels)
- E-Marketing
- Ordering books and CDs over the web (such as www.amazon.com)

- Portal sites (such as Yahoo and Hotmail)
- Recruitment Services (such as www.jobserve.com)
- Internet Banking
- On-line casinos (for gambling)
- Newspapers and News Channels
- On-line shopping and Shopping Malls

The prediction for the growth of e-commerce businesses in the early days of the world-wide web was that the new web-based economy would replace the traditional bricks and mortars companies. The expectation was that most business would be conducted over the web, with traditional enterprises losing market share and eventually going out of business. Exponential growth of the e-commerce companies was predicted, and the size of the new web economy was estimated to be trillions of U.S. dollars.

New companies were formed to exploit the opportunities of the web, and existing companies developed e-business and e-commerce strategies to successfully adapt to the challenge of the brave new world. Companies providing full e-commerce solutions were concerned with the selling of products or services over the web to either businesses or consumers. These business models are referred to as Business-to-Business (B2B) or Business-to-Consumer (B2C). E-commerce web sites have the following characteristics (Table 6.3):

Table 6.3 Characteristics of e-commerce

Feature	Description
Catalogue of products	The catalogue of products details the products available for sale and their prices.
Well designed and easy to use	This is essential as otherwise the web site will not be used.
	Usability requires that it is easy to find the relevant information on the web site and that the performance of the site is good.
Shopping carts	This is analogous to shopping carts in a supermarket. The electronic shopping cart contains the products that the user plans to purchase. The user may add or remove items from the cart.
Security	Security of credit card information is a key concern for users of the web, as users need to have confidence that their credit card details may not be intercepted. There are technologies such as Secure Socket Layer (SSL) that provide encryption of credit card Information. Encryption protects the privacy of the information from un-authorized access.
Payments	Once the user has completed the selection of purchases there is a check-out facility to arrange for the purchase of the goods. Payment for the products is generally by credit or debit card once the user has completed shopping.
Order Fulfillment/Order Enquiry	Once payment has been received the products must be delivered to the customer.

6.7 Dot Com Companies

The growth of the internet and world-wide web was exponential, and the boom led to the formation of many "new economy" businesses. The key characteristic of these new businesses was that business was largely conducted over the web as distinct from the "old economy" bricks and mortar companies. These new companies included the internet portal site "yahoo.com", the on-line book store "amazon.com", and the on-line auction site "ebay.com". Yahoo.com provides news and a range of services, and it earns most of its revenue from advertisements. Amazon.com sells books as well as a collection of consumer and electronic goods. Ebay.com brings buyers and sellers together in an on-line auction space and allows users to bid for items. Boo.com was an on-line fashion company that failed dramatically in late 1999. Pets.com was an on-line pet supplies and accessory company that lasted one year in business. Priceline.com is an on-line travel firm that offers airlines and hotels a service to sell unfilled seats or rooms cheaply. ETrade.com is an on-line share dealing company. Some of these new technology companies were successful and remain in business. Others were financial disasters due to poor business models, poor management and poor implementation of the new technology.

Some of these technology companies offered a replacement or internet version of a traditional bricks and mortar company, whereas others offered a unique business offering. For example, the technology company "ebay.com" offers an auctioneering internet site to consumers worldwide. There was no available service like this available, and the service offering was quite distinct from traditional auctioneering.

The internet portal company "yahoo.com" was founded by David Filo and Jerry Yang who were students at Stanford. It was initially a hobby for the two students, who used it as a way to keep track of their personal interests and their corresponding web sites on the internet. Their list of interests grew over time and became too long and unwieldy. Therefore, they broke their interests into a set of categories and then sub-categories, and this is the core concept of yahoo.com. The term "yahoo" may stand for "Yet another Hierarchical Officious Oracle" or possibly it may have been taken from the name of the primitive creatures in Gulliver Travels by the Irish author Jonathon Swift. There was a lot of interest in the site from other students, family and friends and a growing community of users. The founders realized that the site had commercial potential, and they incorporated it as a business in 1995. The company launched its initial public offering (IPO) one year later in April 1996, and the company was valued at $850 million (or approximately 40 times its annual sales). The headquarters of the company are in California.

Yahoo is a portal site and offers free email accounts to users, a search engine, news, shopping, entertainment, health, and so on. The company earns a lot of its revenue from advertisement (especially the click through advertisements that appear on a yahoo web page). Research indicates that only about 2% of users actually click through on these advertisements, and it is therefore questionable whether these advertisements are actually effective. The company also earns revenue from providing services for businesses. These include web hosting services, web tools, larger mailboxes, personal advertisements for dating, and so on. The company has been very successful.

Amazon is a famous internet company founded by Jeff Bezos in 1995. It was one of the earliest companies to sell products over the internet, and it commenced business as an on-line bookstore. Its product portfolio diversified over time to include the sale of CDs, DVDs, toys, computer software and video games. The name "Amazon" was chosen to represent the volume of products in the company's catalogue, and it was therefore appropriate to name if after the world's largest river. The initial focus of the company was to build up the "Amazon" brand throughout the world. Its goal was to become the world's largest bookstore, and in order to build up market share it initially sold books at a loss by giving discounts to buyers. Its strategy in building the brand through advertisements, marketing and discounts worked very effectively, and the company is now well-established and recording healthy profits.

The company headquarters is in Seattle, Washington. The IPO of the company took place in 1997 and was highly successful. It has been one of the most successful internet companies, and it has become the largest on line bookstore in the world. Many products on the Amazon web site are reviewed to give the buyer an indication as to how good the product is. Amazon has all the essential characteristics for a successful internet company. It has a sound business model with a very large product catalogue; a well-designed web site with good searching facilities; good check out facilities and good order fulfillment. The company has also developed an associates model, and this allows the many associates of Amazon to receive a commission for purchases of products on Amazon made through the associate site. The associates maintain a link to Amazon from their web site, and when a purchaser follows the link, and purchases a product on Amazon they receive a commission from the company.

The on-line auction web site eBay was founded in California by Pierre Omidyar in 1995. Omiyar was a French–American programmer who was born in Paris. He moved to the United States with his parents, and later studied computing. He worked a programmer at Claris prior to setting up eBay. The eBay site brings buyers and sellers together, and it allows buyers to bid for items. The company earns revenue by charging a commission for each transaction, and it is a profitable company. The IPO of eBay took place in 1998 and was highly successful.

Millions of items, for example, computers, furniture, cars and collectibles are listed, bought and sold on eBay every day. Anything that is legal and does not violate the company's terms of service may be bought or sold on the site. The sellers on eBay are individuals and also large international companies who are selling their products and services. A buyer makes a bid for a product or service, and competes against several other bidders. The highest bid is successful, and payment and delivery is then arranged. The revenue earned by eBay is generated in a number of ways. There are fees to list a product and fees that are applied whenever a product is sold. The company is international with a presence in over twenty countries.

There have been a number of strange offerings on eBay. For example, one man offered one of his kidneys for auction as part of the market for human organs. Other unusual cases have been towns that have been offered for sale (as a joke). Any product lists that violate eBay's terms of service are removed from the site as soon as the company is aware of them. The company also has a fraud prevention mechanism which allows buyers and sellers to provide feedback on each other, and to rate each other following the transaction. The feedback may be positive, negative or neutral, and comments may be included as appropriate. This offers a way to help to reduce fraud with unscrupulous sellers or buyers as they will receive negative ratings and comments.

ETrade.com was launched in 1996 and it allows investors the facility to purchase shares electronically, and it also provides financial information to clients. Priceline was founded by Jay Walker, and offered a service to airlines and hotels to sell unfilled seats or rooms cheaply. Priceline was valued at $10 billion at its IPO. This was despite the fact that unlike airlines it had no assets, and was actually selling flights at a loss.

6.7.1 Dot Com Failures

Several of companies formed during the dot com era were successful and remain in business today. Others had appalling business models or poor management and failed in a spectacular fashion. This section considers some of the dot com failures and highlights the reasons for failure.

Webvan.com was an on-line grocery business that was based in California. It delivered products to a customer's home within a 30-minute period of their choosing. The company expanded to several other cities before it went bankrupt in 2001. Most of the failings at Webvan were due to management as the business model itself was fine. None of the management had experience of the supermarket or grocery business, and the company spent far too much on its infrastructure too quickly, and it rapidly ran out of funding. Its investors had advised the company to build up an infrastructure to deliver groceries as quickly as possible, rather than developing partnerships with existing supermarkets. It built warehouses, purchased a fleet of delivery vehicles and top of the range computer infrastructure before it ran out of money.

Boo.com was founded in 1998 by Ernst Malmsten, Kajsa Leander and Patrik Hedel. It was an on-line fashion retailer based in the United Kingdom. The company went bankrupt in 2000 and it succeeded in wasting over $135 million in shareholder funds in less than 3 years. The company web site was poorly designed for its target audience, and it went against many of the accepted usability conventions of the time. The site was designed in the days before broadband, and the designers of the site assumed that its users would employ 56 K modems. The web site included the latest Java and Flash technologies, and it took several minutes to load the first page of the web site for most users for the first released version of the site. Further, the navigation of the web site was inconsistent, and changed as the user moved around the site. The net effect was that users were turned off the site, and despite extensive advertising by the company, users were not inclined to use the site.

Other reasons for failure included poor management and leadership, lack of direction, lack of communication between departments, spirally costs left unchecked, hiring staff and contractors in large numbers leading to crippling pay roll costs. Further, a large number of products were returned by purchasers, and there was no postage charge applied for this service. However, the company incurred a significant cost for covering postage for returns. Eventually, the company went bankrupt in 2000. A description of the formation and collapse of the company is available in the book Boo Hoo [MaP:02]. This book is a software development horror story, and the maturity of the software development practices in place at the company may be judged by the fact that while it had up to 18 contractor companies working to develop the web-site the developers were working without any source code control mechanism in place. The latter is basic software engineering.

Pets.com was an on-line pet supply company founded in 1998 by Greg McLemore. It sold pet accessories and supplies. It had a well-known advertisement as to why one should shop at an online pet store. The answer to this question was: "Because Pets Can't Drive!". Its mascot (the Pets.com sock puppet) was also well known. It launched its IPO in February 2000 just before the dot com collapse.

Pets.com made investments in infrastructure such as warehousing and vehicles. However, in order to break even the company needed a critical mass of customers. The company's management believed that the company needed $300 million of revenue to breakeven, and that this would take a minimum of four to five years to achieve. This time period was based on growth of internet shopping and the percentage of pet owners that shopped on the Internet. By late 2000, the company realised that it would be unable to raise further capital due to the negative sentiment with technology firms following the dot com collapse. They tried to sell the company without success. The company went into liquidation at the end of November 2000 (just nine months after the IPO). However, it did return 17 cents in the dollar to its shareholders.

Kozmo.com was an online company that promised free one-hour delivery of small consumer goods. It was founded by Joseph Park and Yong Kang in New York in 1998. The point-to-point delivery was usually done by bicycle messengers, and no delivery fee was charged. However, the business model was deeply flawed, as the point to point delivery of small goods within a 1-hour period is extremely expensive. There was no way that the company could make a profit unless it charged delivery fees. The company argued that they could make savings to offset the delivery costs as they did not require retail space. The company expanded into several cities in the United States, and raised about $280 million from investors. It had planned to launch an IPO but due to the dot com bubble this was abandoned. The company shut down in 2001.

6.7.2 Business Models

A company requires a sound business model to be successful. A business model converts a technology idea or innovation into a commercial reality, and needs to be appropriate for the company and the market that the company is operating in. A company needs a lot more than a good idea or invention to be successful. A company with an excellent business idea but with a weak business model may fail, whereas a company with an average business idea but an excellent business model may be quite successful. Many of the business models in the dot com era were deeply flawed, and the eventual collapse of many of these companies was predictable. Chesbrough and Rosenbroom [ChR:02] have identified six key components in a business model (Table 6.4):

Table 6.4 Characteristics of business models

Constituent	Description
Value Proposition	This describes how the product or service that the company provides is a solution to a customer problem. The value that the product or service provides is described from the customer's perspective.
Market Segment	This describes the customers that will be targeted, and it may include several market segments. Each segment may have different needs.
Value Chain Structure	This describes where the company fits into the value chain. The value chain [Por:98] includes activities such as inbound logistics for receiving raw materials, operations for converting raw materials to product, outbound logistics for order-fulfillment and delivery to the customer, marketing and sales activities, and customer support and maintenance services. These activities are supported by functions for procurement, technology development, HR, sales and marketing.
Revenue Generation and Margins	This describes how revenue will be generated and includes the various revenue streams (e.g., sales, investment income, support income, subscriptions, and so on). It also describes the cost structure (employee, rent, marketing, and so on) and the target profit margins.
Position in Value Network	This involves identifying competitors and other players that can assist in delivering added value to the customer.
Competitive Strategy	This describes how the company will develop a competitive advantage to be a successful player in the market. For example, the company may plan to be more competitive on cost than other players, or it may be targeting a niche market.

6.7.3 Bubble and Burst

Netscape was founded as Mosaic Communications by Marc Andreessen and Jim Clark in 1994. It was renamed as Netscape in late 1994. The initial public offering of the company in 1995 demonstrated the incredible value of the new internet companies. The company had planned to issue the share price at $14 but decided at the last minute to issue it at $28. The share price reached $75 later that day. This was followed by what became, in effect, the dot com bubble where there were a large number of public offerings of internet stock, and where the value of these stocks reached astronomical levels. Eventually, reality returned to the stock market when it crashed in April 2000, and share values returned to more sustainable levels.

The valuations of these companies were extraordinary. The vast majority of these companies were losing substantial sums of money, and few expected to deliver profits in the short term. Traditional forms of determining the value of a company using accounts such as the balance sheet, profit and loss account, and price to earnings ratio were ignored. Instead, investment bankers argued that there was now a new paradigm in stock valuation. This brave new world argued that the potential future earnings of the stock should be considered in determining the appropriate valuation of the stock. This was used to justify the high prices of shares in technology companies, and frenzied investors rushed to buy these over-priced and over-hyped stocks. Common sense seemed to play no role in decision making, and this is a characteristic of a bubble. The dot com bubble included features such as:

- Irrational exuberance on the part of investors
- Insatiable appetite for Internet Stocks
- Incredible greed from all parties involved
- A lack of rationality by all concerned
- No relationship between Balance Sheet/Profit and Loss and the share price
- Interest in Making Money rather than in building the business first
- Following Herd Mentality
- Questionable Decisions by Federal Reserve Chairman Alan Greenspan
- Questionable Analysis by Investment Firms
- Conflict of Interest in Investment Banks between Analysis for Investors and earning revenue from IPOs
- Market had left reality behind

However, many individuals and companies made a lot of money from the boom, although many other investors including pension funds and life assurance funds made significant losses. The investment banks typically earned 5%–7% commission on each successful IPO, and it was therefore in their interest not to question the boom too closely. Those investors that bought early and disposed early obtained a good return; however, many of those who held shares for too long incurred losses.

The full extent of the boom can be seen from Figs. 6.3 and 6.4 which show the increase in value of the Dow Jones and Nasdaq from 1995 through 2002.

The extraordinary rise of the Dow Jones from a level of 3800 in 1995 to 11900 in 2000 represented a 200% increase over 5 years or approximately 26% annual growth (compound) over 5 years. The rise of the Nasdaq over this time period is even more dramatic. It rose from a level of 751 in 1995 to 5000 in 2000 representing a 566% increase over the time period. This is equivalent to a 46% compounded annual growth rate of the index.

The fall of the indices have been equally dramatic especially in the case of the Nasdaq which peaked at 5000 in March 2000 had fallen by 76%–1200% by September 2002. It had become clear that internet companies were rapidly going through

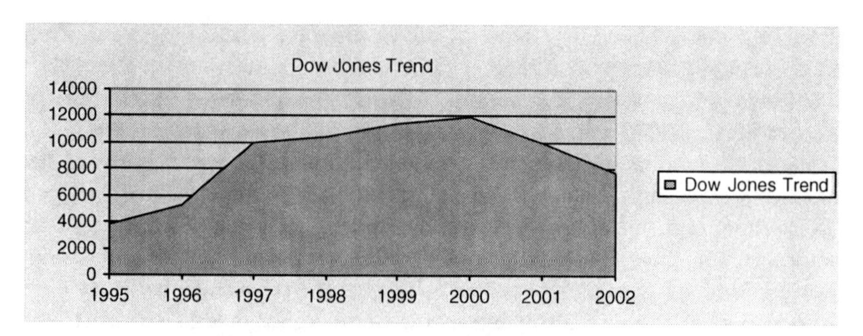

Fig. 6.3 Dow Jones (1995–2002)

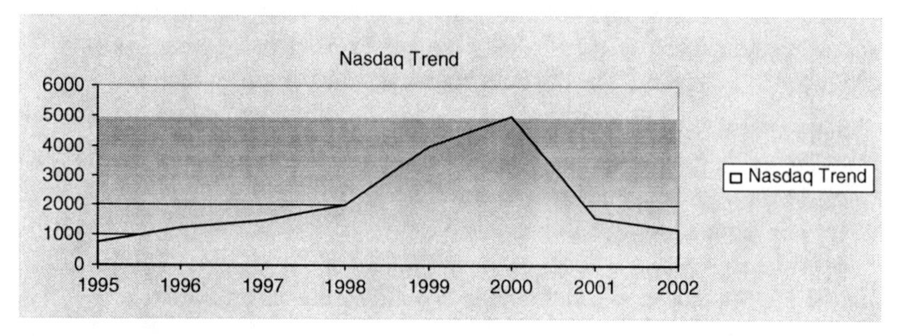

Fig. 6.4 Nasdaq (1995–2002)

the cash raised at the IPOs, and analysts noted that a significant number would be out of cash by end of 2000. Therefore, these companies would either go out of business, or would need to go back to the market for further funding. This led to questioning of the hitherto relatively unquestioned business models of many of these internet firms. Funding is easy to obtain when stock prices are rising at a rapid rate. However, when prices are static or falling, with negligible or negative business return to the investor, then funding dries up. The actions of the Federal Reserve in rising interest rates to prevent inflationary pressures also helped to correct the irrational exuberance. However, it would have been more appropriate to have taken this action 2–3 years earlier.

Some independent commentators had recognized the bubble but their comments and analysis had been largely ignored. These included "The Financial Times" and the "Economist" as well as some commentators in the investment banks. Investors rarely queried the upbeat analysis coming from Wall Street, and seemed to believe that the boom would never end. They seemed to believe that rising stock prices would be a permanent feature of the United States stock markets. Greenspan had argued that it is difficult to predict a bubble until after the event, and that even if the bubble had been identified it could not have been corrected without causing a contraction. Instead, the responsibility of the Fed according to Greenspan was to mitigate the fallout when it occurs.

There have, of course, been other stock market bubbles throughout history. For example, in the 1800s there was a rush on railway stock in England leading to a bubble and eventual burst of railway stock prices in the 1840s.

6.8 E-Software Development

The growth of the world-wide web and electronic commerce in recent years has made the quality, reliability and usability of the web sites a key concern. An organisation that conducts part or all of its business over the world-wide web will need to ensure that its web site is fit for purpose. Software development for web-based systems is relatively new, and an introduction is provided in [Nie:01]. Web applications are quite distinct from other software systems in that:

- They may be accessed from anywhere in the world.
- They may be accessed by many different browsers.
- The world-wide web is a distributed system with millions of servers and billions of users.
- The usability and look and feel of the application is a key concern.
- The performance of the web site is a key concern.
- Security threats may occur from anywhere.
- The web site must be capable of dealing with a large number of transactions at any time.
- The web site has very strict availability constraints (typically $24 \times 7 \times 365$ availability).
- The web site needs to be highly reliable.

There are various roles involved in web based software development including content providers who are responsible for providing the content on the web; web designers who are responsible for graphic design of the web site; programmers who are responsible for the implementation; and administrators who are responsible for administration of the web site. These roles may be combined in practice.

Rapid Application Development (RAD) or Joint Application Development (JAD) lifecycles are often employed for web site development. The lifecycle followed is usually spiral rather than waterfall, and it is often the case that the requirements for web-based systems will not be fully defined at project initiation. This is especially the case with usability requirements that generally evolve to the final agreed set. It is generally inappropriate to employ the waterfall lifecycle for this domain.

The software development proceeds in a number of spirals where each spiral typically involves updates to the requirements, design, code, testing, and a user review of the particular iteration or spiral.

The spiral is, in effect, a re-usable prototype and the customer examines the current iteration and provides feedback to the development team to be included in the next spiral. The approach is to partially implement the system. This leads to a better understanding of the requirements of the system and it then feeds into the next

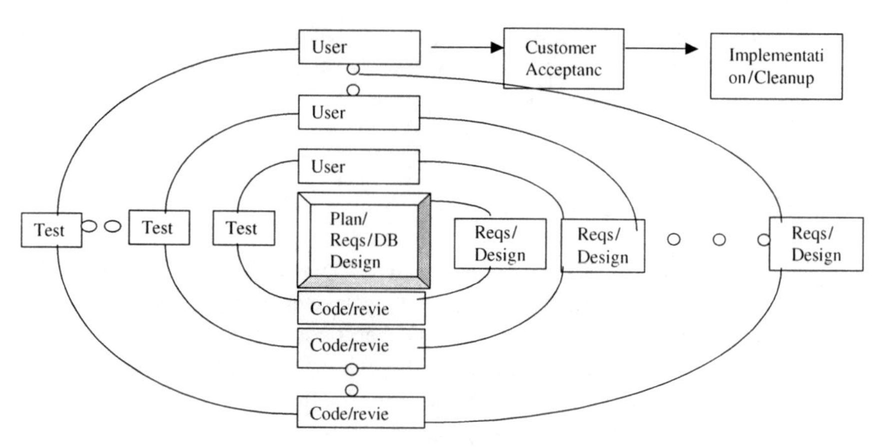

Fig. 6.5 Spiral lifecycle model

cycle in the spiral. The process repeats until the requirements and product are fully complete. The spiral model is shown in Fig. 6.5.

Web site quality and performance is a key concern to companies and their customers. The implementation of a web site requires sound software engineering practices to design, develop and test the web site.

A spiral lifecycle is usually employed and the project team needs to produce similar documentation as the waterfall model, except that the chronological sequence of delivery of the documentation is more flexible. The joint application development is important as it allows early user feedback to be received on the look and feel and correctness of the application. The approach is often "design a little, implement a little, and test a little" as this is generally the best approach to developing a usable application.

Various technologies are employed in web developed. These include HTML which is used to design simple web pages and was originally developed by Berners-Lee; CGIs (Common Gateway Interface) are often employed in sending a completed form to a server; cookies are employed to enable the server to store client specific information on client's machine. Other popular technologies are Java, Javascript, VB Script, Active X and Flash. There are tools such as Dreamweaver to assist with Web site design.

Testing plays an important role in assuring quality and various types of web testing include:

- Static testing
- Unit testing
- Functional testing
- Browser compatibility testing
- Usability testing
- Security testing
- Load/performance/stress testing

- Availability testing
- Post deployment testing

Static testing generally involves inspections and reviews of documentation. The purpose of static testing of web sites is to check the content of the web pages for accuracy, consistency, correctness, and usability, and also to identify any syntax errors or anomalies in the HTML. The purpose of unit testing is to verify that the content of the web pages correspond to the design, that the content is correct, that all the links are valid, and that the web navigation operates correctly. The purpose of functional testing is to verify that the functional requirements are satisfied. Functional testing may be quite extensive as e-commerce applications can be quite complex. It may involve product catalogue searches, order processing, credit checking and payment processing, as well as interfaces to legacy systems. Also, testing of cookies, whether enabled or disabled, needs to be considered.

The purpose of browser compatibility testing is to verify that the web browsers that are to be supported are actually supported. Different browsers implement HTML differently; for example, there are differences between the implementation by Netscape and Microsoft. These days internet explorer is the main web browser as it now dominates the market.

The purpose of usability testing is to verify that the look and feel of the application is good. The purpose of security testing is to ensure that the web site is secure. The purpose of load, performance and stress testing is to ensure that the performance of the system is within the defined parameters. There are tools to measure web server and operating system parameters and to maintain statistics. These tools allow simulation of a large number of users at one time or simulation of the sustained use of a web site over a long period of time.

There is a relationship between the performance of a system and its usability. Usability testing includes testing to verify that the look and feel of the application are good, and that performance of loading of web pages, graphics, etc., is good. There are automated browsing tools which go through all of the links on a page, attempt to load each link, and produce a report including the timing for loading an object or page at various modem speeds. Good usability requires attention to usability in design, and usability engineering is important for web based or GUI applications.

The purpose of post-deployment testing is to ensure that the performance of the web site remains good, and this is generally conducted as part of a service level agreement (SLA). Service level agreements typically include a penalty clause if the availability of the system or its performance falls below defined parameters. Consequently, it is important to identify as early as possible potential performance and availability issues before they become a problem. Thus post-deployment testing will include monitoring of web site availability, performance, security, etc., and taking corrective action as appropriate. Most web sites are operating $24 \times 7 \times 365$ days a year, and there is the potential for major financial loss in the case of an outage of the electronic commerce web site. There is a good account of e-business testing and all the associated issues by Paul Gerrard in detail in [Ger:02].

6.9 E-Commerce Security

The world-wide web extends over computers in virtually every country in the world. It consists of unknown users with un-predictable behavior operating in unknown countries. These users and web sites may be friendly or hostile and the issue of trust arises.

- Is the other person whom they claim to be?
- Can the other person be relied upon to deliver the goods on-payment?
- Can the other person be trusted not to inflict malicious damage?
- Is financial information kept confidential on the server?

Hostility may manifest itself in various acts of destruction. For example, malicious software may attempt to format the hard disk of the local machine and if successful all local data will deleted. Other malicious software may attempt to steal confidential data from the local machine including bank account or credit card details. One particularly destructive virus is the denial of service attack: this is where a web site is overloaded by a malicious attack, and where users are therefore unable to access the web site for an extended period of time.

The display of web pages on the local client machine may involve the downloading of programs from the server, and running the program on the client machine (e.g., Java Applets). Standard HTML allows the static presentation of a web page, whereas many web pages include active content (e.g., Java Applets or Active X). There is a danger that a Trojan horse[4] may be activated during the execution of active content.

Security threats may be from anywhere (e.g., client side, server side, transmission) in an e-commerce environment, and therefore a holistic approach to security is required. Internal and external security measures need to be considered. Internal security is generally implemented via procedures and access privileges.

It is essential that the user is confident in the security provided as otherwise they will be reluctant to pass credit card details over the web for purchases. This has led to technologies such as secure-socket layer (SSL) and secure HTTP (S-HTTP) to ensure security. Another approach is the use of e-cash for electronic payments.

[4] The origin of the term "Trojan Horse" is from the approach used by the Greeks to capture the city of Troy. The Greek hero Odysseus and others hid in a wooden horse while the other Greeks sailed away from Troy. This led the Trojans to believe that the Greeks had abandoned their attack and were returning to their homeland leading behind a farewell gift for the citizens of Troy. The Trojans brought the wooden horse into the city and later that night Odysseus and his companions opened the gates of Troy to the returning Greeks, and the mass slaughter of the citizens of Troy commenced. Troy is located at the mouth of the Dardanelles in Turkey.

6.10 Review Questions

1. Describe the development of the internet.
2. Describe the development of the world-wide web and its key constituents.
3. Describe the applications of the world-wide web.
4. Describe the key constituents of an electronic commerce site.
5. Describe a successful dot com company that you are familiar with. What has made the company successful?
6. Describe a dot com failure that you are familiar with. What caused the company to fail?
7. Discuss the key components of a business model.
8. Discuss software development in a web environment.

6.11 Summary

The invention of the world-wide web by Tim Berners-Lee was a revolutionary milestone in computing. It has transformed the internet from mainly academic use (with limited commercial involvement) to the world of electronic commerce where the use of the internet and world-wide web is an integral part of peoples' lives.

The growth of the world-wide web has been phenomenal since its invention in 1991. New companies were formed to exploit the commercial opportunities of the web, and existing companies developed e-business and e-commerce strategies. Companies providing full e-commerce solutions were concerned with the selling of products or services over the web to either businesses or consumers. The world-wide web is revolutionary in that:

- No single organization controls the web.
- No single computer controls the web.
- Millions of computers are interconnected.
- It is an enormous market place of millions (billions) of users.
- The web is not located in one physical location.
- The web is a space and not a physical thing.

The growth of the world-wide web was exponential, and the boom led to the formation of many "new economy" businesses. The key characteristic of these new businesses was that business was largely conducted over the web as distinct from the "old economy" bricks and mortar companies. Some of these companies were very successful and remain in business. Others were financial disasters due to poor business models, poor management and poor implementation of the new technology.

Chapter 7
Famous Technology Companies

Key Topics

Hollerith's Tabulating Machine
IBM 360
IBM Personal Computer
Microsoft
Motorola
Six Sigma

7.1 Introduction

This chapter considers the history of some well known technology companies. These include International Business Machines (IBM), Microsoft and Motorola. The origin of IBM goes back to the late nineteenth century with Hermann Hollerith's work on tabulating machines. This machine was designed to tabulate the results of the 1890 census in the United States. IBM has grown to become a major corporation and it has made a major contribution to the computing field. It is a world class professional company and is dedicated to quality and customer satisfaction.

Microsoft was founded in the mid-1970s and it was awarded a contract by IBM to develop the Disk Operating System (DOS) for the IBM PC. It has grown to become a major corporation and has developed operating systems such as Microsoft Windows NT, Windows 2000 and Windows XP.

Motorola was founded as the Galvin Manufacturing Corporation in 1928. Its initial business was the production of radios for cars and it became a world leader in radio and telecommunications. Motorola produces mobile phones and base stations for the mobile telecommunications field.

Apple was founded in the mid-1970s. It developed the Apple Macintosh in the mid-1980s. This introduced a friendly graphical user interface (GUI) which made the machine easy to use. The company also produced the iPod.

Other companies discussed in this chapter include HP, Oracle and Siemens.

7.2 International Business Machines

IBM is a household name today and the company has a long and distinguished history. In many ways, the history of computing is closely related to the history of IBM, as the company has played a key role in the development of the computers that we are familiar with today. The origins of the company go back to the late nineteenth century with the processing of the 1880 census of the United States.

The processing of the 1880 census took 7 years to complete as all the processing was done manually. The population of the United States was steadily increasing at this time, and it was predicted that the 1890 census would take in excess of 10 years to process. The US Census Bureau recognised that the current methodology for processing census results was no longer fit for purpose. Therefore, it decided to hold a contest among its employees in an effort to find a more efficient methodology to tabulate the census data. The winner was Hermann Hollerith who was the son of a German immigrant (Fig. 7.1).

Fig. 7.1 Hermann Hollerith
Courtesy of IBM archives.

Fig. 7.2 Hollerith's Tabulator (1890)
Courtesy of IBM archives.

His punch card Tabulating Machine used an electric current to sense holes in punched cards, and it kept a running total of the data (Fig. 7.2). His machine used Babbage's idea of punched cards for data storage. The new methodology enabled the results of the 1890 census to be available within 6 weeks, and the population was recorded to be over 62 million.

7.2.1 Early Years

Hollerith formed the Tabulating Machine Company in Washington, D.C., in 1896. This was the world's first electric tabulating-machine company, and it later merged with the International Time Recording Company to form the Computing Tabulating Recording Company (CTR) in 1911. Thomas Watson Senior became president of the Computing Tabulating Company in 1914 and at that time the company employed 400 people. Watson transformed the company over the following decades to become a global multinational. He was responsible for the famous "Think" signs that have been associated with IBM for many years. They were first used in the company in 1915.

Watson considered the motivation of the sales force to be an essential part of his job, and the sales people were required to attend an IBM sing along. The verses in the songs were in praise of IBM and its founder Thomas Watson Sr (Fig. 7.3). The company was renamed in 1924 to become International Business Machines (IBM). It employed over 3,000 people in 1924, and had revenues of US $11 million, with

Fig. 7.3 Thomas Watson Sr
Courtesy of IBM archives.

a net income of $2 million. It had manufacturing plants in the United States and Europe. By 2005, IBM had revenues of $91.1 billion, net income of $7.9 billion, and it employed over 300,000 people.

IBM has evolved to adapt its business to a changing world and emerging trends in the information handling field. Its early products were designed to process, store and retrieve information from tabulators and time recording clocks. Today, the company produces powerful computers and global networks, and it is an industry leader in the development of computer systems to meet the needs of business and consumers.

IBM redesigned the existing punch card in the late 1920s to create the popular IBM punch card. It had 80 columns which was almost double the capacity of existing cards. The new design was patented, and it included rectangular holes. The company introduced a mechanism by which staff could make improvement suggestions in the 1920s. Many companies today now encourage their staff to make improvement suggestions to deliver better results.

The great depression of the 1930s affected many American companies and had a devastating impact on the lives of many Americans. However, surprisingly, its impact on IBM was minimal, and the company continued to grow. The policy of Thomas Watson Sr. and IBM was to take care of its employees, and IBM was one of the first corporations to provide life insurance and paid vacations for its employees. Watson kept his workers busy during the depression by producing new machines even while demand was slack. He also won a major government contract to maintain employment records for over 26 million people.

Watson created a division in the early 1930s to lead the engineering, research and development efforts for the entire IBM product line. This was followed by the creation of a modern research and development laboratory, and also a centre dedicated to education and training in New York. The education and development of employees was seen as essential to the success of the business. IBM launched an employee and customer magazine called *Think* in the 1930s. This magazine included topics such as education and science.

IBM contributed to the defense of the United States during the Second World War. It placed all of its plants at the disposal of the US government, and it expanded its product line to include military equipment such as rifles. IBM commenced work on computing during the war years. This included work on the Harvard Mark I (also known as ASCC) machine completed in 1944. It was essentially an electromechanical calculator that could perform large computations automatically.

The machine was 50 feet long and eight feet high and weighed five tons. It performed additions in less than a second, multiplications in 6 seconds, and division in about 12 seconds. It used electromechanical relays to perform the calculations.

7.2.2 Early IBM Computers

The company developed the Vacuum Tube Multiplier in 1943 and this was important in the move from electromechanical to electronic machines. It was the first complete machine ever to perform arithmetic electronically by substituting vacuum tubes for electric relays. The key advantages of the vacuum tubes is that they were faster, smaller, and easier to replace than the electromechanical switches used in the Mark I. They allowed engineers to process information thousands of times faster.

The company introduced its first large computer based on the vacuum tube in 1952. This machine was called the IBM 701, and it executed 17,000 instructions per second (Fig. 7.4). It was used mainly for government work and also for business applications. Transistors began to replace vacuum tubes in the late 1950s. Thomas Watson, Sr., retired in 1952 and his son, Thomas Watson, Jr., became chief executive officer the same year.

Thomas Watson, Jr. believed that the future of IBM was in computers rather than in tabulators (Fig. 7.5). He recognized the future role that computers would play in business, and realised that IBM needed to change to adopt to the new technology world. He played a key role in the transformation of IBM from its existing business model to a company that would become the world leader in the computer industry.

IBM introduced the IBM 650 (Magnetic Drum Calculator) in 1954. This was an intermediate sized electronic computer designed to handle widely diversified accounting and scientific computations. It was used by universities and businesses. The machine included a central processing unit, a power unit and a card reader.

Fig. 7.4 IBM 701
Courtesy of IBM archives.

The IBM 704 data processing system was a large computer that was introduced in 1954. It included core memory and floating-point arithmetic, and was used for both scientific and commercial applications. It included high speed memory which was faster and much more reliable than the cathode-ray-tube memory storage mechanism employed in earlier machines. It also had a magnetic drum storage unit which could store parts of the program and intermediate results.

The interaction with the system was either by magnetic tape or punched cards entered through the card reader. The program instructions or data were initially produced on punched cards. They were then either converted to magnetic tape or read directly into the system, and data processing performed. The output from the processing was then sent to a line printer, magnetic tape or punched cards. Multiplication and division was performed in 240 microseconds.

The designers of the IBM 704 included John Backus and Gene Amdahl. Backus was one of the key designers of the Fortran programming language introduced by IBM in 1957. Fortran was the first scientific programming language and has been used extensively by engineers and scientists. Gene Amdahl later went on to found his own company in 1970 (i.e., the Amdahl Corporation[1]).

The first commercial computer with transistor logic (the IBM 7090) was introduced in 1958. It was designed for large-scale scientific applications, and it was over 13 times faster than the older vacuum tube IBM 701. It could perform 229,000

[1] Amdahl went on to rival IBM and became a threat to the success of IBM. At its peak, Amdahl had 22% of the mainframe market. The company bacame a wholly owned subsidiary of Fujitsu in 1997, and it is located in San Francisco.

Fig. 7.5 Thomas Watson Jr
Courtesy of IBM archives.

calculations per second. It used a 36-bit word and had an address-space of 32,768 words. It was used by the US Air Force to provide an early warning system for missiles and also by NASA to control space flights. It cost approximately $3 million but it could be rented for over $60 K per month.

IBM introduced the first computer disk storage system in 1957. This medium was called the Random Access Method of Accounting and Control (RAMAC), and it became the basic storage medium for transaction processing. The RAMAC's random access arm could retrieve data stored on any of the 50 spinning disks.

IBM contributed to the Semi-Automatic Ground Environment (SAGE) early warning system during the cold war. The US Air Force had commenced work on SAGE in 1954, and IBM provided the hardware for the system. The initial installation was completed in 1958, and the system was fully implemented in 1963. It remained operational until 1984.

There were 24 SAGE Direction Centers and three SAGE Combat Centers located in the US. Each centre was linked by long-distance telephone lines, and Burroughs provided the communications equipment that allowed the centers to communicate with one another. This was one of the earliest computer networks. Each center contained a large digital computer that automated the information flow, and provided real time control information on aircraft and on weapons systems. It tracked and identified aircraft, and presented the electronic information to operators on a display device (cathode ray tub).

The IBM 1401 data processing system and the IBM 1403 printer were launched in 1959. The 1401 was an all-transitorised data processing system and it was aimed at small businesses. It included high speed card punching and reading, magnetic tape input and output and high speed printing. The 1403 printer was four times faster than any competitor printer. IBM introduced a program termed "Speak Up" to enhance staff communication in 1959. It opened the headquarters for its research division at York town Heights, New York in 1961.

7.2.3 The IBM System 360

Thomas Watson announced the new System 360 to the world at a press conference in 1964 and said:

The System/360 represents a sharp departure from concepts of the past in designing and building computers. It is the product of an international effort in IBM's laboratories and plants and is the first time IBM has redesigned the basic internal architecture of its computers in a decade. The result will be more computer productivity at lower cost than ever before. This is the beginning of a new generation – not only of computers – but of their application in business, science and government.

The chief architect for the 360 was Gene Amhadl and the S360 project manager was Fred Brooks.[2]

The IBM 360 was a family of small to large computers, and it offered a choice of five processors and 19 combinations of power, speed and memory (Fig. 7.6). There were 14 models in the family. The concept of a "family of computers" was a paradigm shift away from the traditional "one size fits all" philosophy of the computer industry, as up until then, every computer model was designed independently. The family of computers ranged from minicomputers with 24 KB of memory, to supercomputers for US missile defence systems. However, all these computers had the same user instruction set, and the main difference was that the larger computers implemented the more complex machine instructions with hardware, whereas the smaller machines used microcode.

The S/360 Architecture potentially allowed customers to commence with a lower cost computer model, and to then upgrade over time to a larger system to meet their evolving needs. The fact that the same instruction set was employed meant that the time and expense of re-writing software was avoided. The machines had

[2] Fred Brooks is the author of "The Mythical Man Month". This is a well-known publication that considers the challenges of delivering a major project (of which software is a key constituent) on time, on budget and with the right quality.

Fig. 7.6 IBM 360 Model 30
Courtesy of IBM archives.

different operating systems, with the smaller machines having a very basic operating system, the mid-range using an operating system called DOS, and the larger systems using an operating system called OS/360. The System 360 also gave significant performance gains, as it included IBM-designed solid logic technology. This allows denser and faster circuits than earlier transistors, and was over 1,000 times more reliable than vacuum tubes.

The S/360 was used extensively in the Apollo project to place man on the moon. The contribution by IBM computers and personnel were essential to the success of the project. IBM invested over $5 billion in the design and development of the S/360. However, the gamble paid off and it was a very successful product line. The System/360 introduced a number of new industry standards including:

- 8-bit Bytes
- Byte Addressable Memory
- 32-bit words
- Two's complement Arithmetic
- EBCDIC Character Set
- Microcoded CPUs
- IBM Floating Point Architecture.

IBM introduced the Customer Information Control System (CICS) in 1969. This is a transaction processing system designed for online and batch processing. It was originally developed at IBM's Palo Alto laboratory, but development moved to IBM's laboratory at Hursley, England from the mid-1970s. It is still developed and enhanced at Hursley.

It is used by banks and insurance companies in the financial sector for their core business functions. It can support several thousand transactions per second, and up to 300 billion transactions flow through CICS every day. It is available on large mainframes and on several operating systems, including Z/OS, AIX, Windows and Linux. CICS applications have been written in Cobol, PL/1, C and Java.

The IBM System/370 was introduced in 1970. It was backwards compatible with the older 360 system, in that programs that ran on the 360 could still run on the 370. This made it easier for customers to upgrade from their System 360 to the System 370. The S/370 employed Virtual Memory.[3]

The floppy disk was introduced in 1971, and it became the standard for storing personal computer data. The IBM 3340 Winchester disk drive was introduced in 1973. It doubled the information density on disk surfaces and included a small light read/write head that was designed to ride on an air film that was 18×10^{-6} inches thick. Winchester technology was employed up to the early 1990s.

IBM introduced the Systems Network Architecture (SNA) networking protocol in 1974, and this protocol provided a set of rules and procedures for communication between computers. It remained an important standard until the open architecture standards appeared in the 1990s.

It introduced the IBM 5100 Portable Computer in 1975 which cost under $20,000. This was a desktop machine and was used by engineers and scientists. IBM's Federal Systems Division built the flight computers and special hardware for the space-shuttle program.

IBM developed the Data Encryption Standard (DES) in the mid-1970s. DES provides a high degree of security during the transmission of data over communication channels. It specifies an algorithm that enciphers and deciphers a message. The effect of enciphering is to make the message meaningless to unauthorized discovery as the task of breaking the encryption algorithm is extremely difficult.

7.2.4 The IBM Personal Computer

IBM introduced the IBM Personal Computer (or PC) in 1981 and this machine was intended to be used by small businesses and in the home (Fig. 7.7). Its price was $1,565 and it was the lowest price computer produced up to that date. It offered 16 kilobytes of memory (that was expandable to 256 kilobytes), a floppy disk and a monitor. The IBM Personal Computer became an immediate success and became the industry standard. The personal computer also led to a new industry of "IBM-compatible" computers. These compatible computers had all of the essential features of the IBM PC but were typically retailed for a much lower price. IBM introduced the IBM Personal Computer XT in 1983. This model had more memory, a dual-sided diskette drive and a high-performance fixed-disk drive. The Personal Computer/AT was introduced by IBM in 1984.

[3] Virtual Memory was developed for the Atlas Computer at Manchester University in England in the early 1960s. It allowed the actual memory space of a computer to appear much larger by using the space available on the hard drive. The Atlas computer was a joint venture between the Manchester University, Ferranti and Plessey.

Fig. 7.7 IBM Personal Computer
Courtesy of IBM archives.

IBM had traditionally produced all of the components for its machines. However, it outsourced the production of components to other companies for the IBM PC. The production of the processor chip was outsourced to a company called Intel,[4] and the development of the Disk Operating System (DOS) was outsourced to a small company called Microsoft.[5] These two companies would later become technology giants.

The IBM token-ring local area network was introduced in 1985. This enabled personal computer users to share printers, files and information within a building. It is essentially a computer network in which all of the computers are arranged in a circle (or ring). There is a special data frame termed a token, and the token moves from computer to the next computer until it arrives at a computer that needs to transmit data. This computer then converts the token frame into a data frame for transmission. That is, the computer that wishes to transmit catches the token,

[4] Intel was founded by Bob Noyce and Gordon Moore in 1968.
[5] Microsoft was founded by Bill Gates and Paul Allen in 1975.

attaches a message to it, and then sends it around the network. The Token Ring network later became the IEEE 802.5 standard.

The Ethernet local area network was developed by Robert Metcalfe at Xerox and its performance was superiour to the IBM Token Ring network. Ethernet was first published as a standard in 1980, and it was later published as the IEEE 802.2 standard. Metcalfe formed the technology company 3-Com to exploit the Ethernet technology. IBM introduced the Advanced Peer-To-Peer Networking architecture (APPN) in 1984. This was widely used for communication by mid-range systems and it allowed individual computers to talk to one another without a central server.

IBM developed the Reduced Instruction Set Computer (RISC) architecture. This technology boosts computer speed by using simplified machine instructions for frequently used functions. It reduces the time to execute commands and is the basis of most workstations in use today. Early work on RISC architecture goes back to work done by IBM in the mid-1960s. Later work by IBM led to the design of the RS/6000 and the subsequent development of the Power PC architecture. The RISC System/6000 was introduced in 1990. It is a family of workstations that were among the fastest and most powerful in the industry. IBM introduced the next generation of personal computers termed the Personal System/2 (PS/2) in 1987. It included a new operating system called Operating System/2 (OS/2). The latter gave users of personal computers access to multiple applications, very large programs and data, and allowed concurrent communication with other systems. It was the first offering in IBM's Systems Application Architecture (SAA) which was designed to make application programs look and work in the same manner across different systems such as personal computers, mid-range systems and larger systems.

A research group at IBM developed a suite of Antivirus tools to protect personal computers from attacks from viruses. This led to the establishment of the High Integrity Computing Laboratory (HICL) at IBM. This laboratory went on to pioneer the science of computer viruses.

IBM researchers introduced very fast computer memory chips in 1988. These chips could retrieve a single bit of information in 2×10^{-8} of a second. This was over four times faster than the existing generation of dynamic random access memory (DRAM) chips. IBM also introduced the IBM Application System/400 (AS/400) in 1988. This was a new family of easy-to-use computers designed for small and intermediate-sized companies. It became one of the world's most popular business computing systems.

A team of IBM researchers succeeded in storing a billion bits of information (i.e., a gigabit) on a single square inch of disk space in 1989. This was 15–30 times greater than the existing data density on magnetic storage devices. The amount of data that a gigabit can store is equivalent to approximately 100,000 A4 pages. IBM introduced a laptop computer in 1991 to give customers computing capabilities on the road or in the air.

IBM, Apple Computers and Motorola entered an agreement in 1991 to link Apple computers to IBM networks, and to develop a new reduced instruction set microprocessors for personal computers. IBM and Motorola completed development and fabrication of the PowerPC 620 microprocessor in 1994. The new open-systems environment allowed both IBM AIX and Macintosh software programs to run on RISC-based systems from both companies.

The introduction of the personal computer represented a paradigm shift in computing, and it led to a fundamental change in the way in which people worked. It placed the power of the computer directly in the hands of millions of people. The previous paradigm was that an individual user had limited control over a computer, and the access privileges of the individual users were strictly controlled by the system administrators. The subsequent introduction of the client–server architecture led to the linking of the personal computers (clients) to larger computers (servers). These servers contained large amounts of data that could be shared with the individual client computers.

IBM had until then provided a complete business solution to its clients with generally one key business person making the decision to purchase the IBM computer system for the company. The personal computer market and the client–server architecture had now fragmented this traditional market, as departments and individuals could now make their own purchasing decisions. The traditional customer relationship that IBM had with its clients had been fundamentally altered. It took IBM some time to adjust to this changing world, and it incurred huge losses of over $8 billion in 1993. The company embarked on cost cutting measures as it worked to adapt to the new environment. This involved reducing its work force, rebuilding IBM's product line, and major cost reductions. IBM's strength in providing integrated business solutions proved to be an asset in adapting to the brave new world.

IBM faced further challenges to adapt to the rise of the Internet and to network computing. The internet was another dramatic shift in the computing industry, but IBM was better prepared this time after its painful adjustment in the client/server market. IBM's leadership helped to create the e-business revolution, and IBM actually coined the term "e-business". IBM outlined to customers and to its employees how the Internet had the ability to challenge older business models, and to transform the nature of transactions between businesses and individuals.

IBM created the world's fastest and most powerful general purpose computer in 1994. This was a massively parallel computer capable of performing 136 billion calculations per second. The increase in computational power of computers was becoming phenomenal. The Deep Blue computer programmed chess program defeated Garry Kasparov in 1997 (Fig. 7.8). Kasparov was then the existing world champion in chess, and the IBM victory showed that the computational power of computers could match or exceed that of man. It was also the first time that a computer had defeated a top-ranked chess player in tournament play. Deep Blue had phenomenal calculating power, and it could calculate 200 million chess moves per second.

IBM and the US Energy Department introduced Blue Pacific which was the world's fastest computer in 1998. It was capable of performing 3.9 trillion calculations per second and had over 2.6 trillion bytes of memory. An indication of its computability is given by the fact that the amount of calculations that this machine could perform in one second would take a person using a calculator over 63,000 years.

Fig. 7.8 Deep Blue Processors
Courtesy of IBM archives.

The Year-2000 millennium bug generated significant customer demand in the late 1990s, as customers wished to ensure that their software was compliant and would continue to function correctly in the new millennium. January 1, 2000 went smoothly for most companies as they had converted their legacy software systems to be Year-2000 compliant.

IBM is a highly innovative company and is awarded more patents[6] in the United States than any other company. It earned over 3,000 patents in 2004. The company is a household name and it has a long and distinguished history. The history of computing is, in many ways, closely related to the history of IBM, as the company has played a key role in the development of the computers that we are familiar with today.

7.3 Microsoft

Microsoft was created by Bill Gates and Paul Allen in 1975 (Fig. 7.9). Steve Ballmer joined the company in 1980. The first real success of the company was with the Disk Operating System (DOS) for personal computers. IBM had originally intended awarding the contract for the operating system to Digital Research use a version of Digital's CP/M operating system on the forthcoming IBM Personal Computer. However, negotiations between IBM and Digital failed in 1981, and IBM awarded the contract to Microsoft to produce a version of the CP/M operating system for its personal computers.

[6] A patent is legal protection that is given to an inventor, and allows the inventor to exploit the invention for a fixed period of time (typically 20 years).

Fig. 7.9 Bill Gates
Photo courtesy of Wikipedia.

Microsoft purchased a CP/M clone called QDOS to assist with this, and IBM renamed the new operating system to PC-DOS. Microsoft created its own version of the operating system called MS-DOS, and the deal with IBM allowed Microsoft to have control of its own QDOS derivative. This proved to be a major mistake by IBM, as MS-DOS became popular in Europe, Japan and South America. The flood of PC clones on the market allowed Microsoft to gain major market share through aggressive marketing of its operating system to the various manufacturers of the cloned PCs. This led to Microsoft becoming a major player in the personal computer market.

The company released its first version of Microsoft Word in 1983, and this would later become the world's most popular word processing package. Microsoft released its first version of Microsoft Windows in 1985, and this product was originally a graphical extension for its MS-DOS operating system. However, later that year, Microsoft and IBM commenced work on a new operating system called Operating System 2 (OS/2). This operating system was for the new IBM PS/2 personal computer. Microsoft introduced its integrated office products called Microsoft Works in 1987, and this product suite Included a word processor, spreadsheet, database and other office applications.

The company introduced its well-known Microsoft Office product in 1989, and this includes Microsoft Word, Microsoft Excel, and Powerpoint. Microsoft introduced Windows 3.0 in 1990, and this new operating system included graphical user interfaces. The company discontinued its work on OS/2, and focused instead on improving Microsoft Windows. Windows (and its successors) became the most popular operating systems in the coming years. Microsoft's office suite gradually became the dominant office suite with a far greater market share than its competitors such as WordPerfect and Lotus 1-2-3. Microsoft released its Windows 3.1 operating system and its Access database software in 1992. Windows was the most widely used GUI operating system by 1993. Windows 95 was released in 1995, Windows NT in 1996, Windows 2000 in 2000, and Windows XP in 2001.

7.3.1 Microsoft Windows and Apple GUI

Apple Computers had taken a copyright infringement lawsuit against Microsoft in 1988, and the legal arguments lasted 5 years. The final ruling in 1993 was in favour of Microsoft. Apple had sought to prevent Microsoft from using GUI elements that were similar to those in Apple's operating system. However, Apple lost the lawsuit and the subsequent appeal. It had claimed that the look and feel of the Macintosh operating system was protected by copyright including 189 GUI elements. However, the judge found that 179 of these had already been licensed to Microsoft (as part of the Windows 1.0 agreement), and that most of the 10 other GUI elements were not copyrightable.

The Apple Computer vs. Microsoft case generated a lot of interest in the computer science community. Some observers considered Apple to be the villain, as they were using legal means to dominate the GUI market and restrict the use of an idea that was of benefit to the wider community. Others considered Microsoft to be the villain with their theft of Apple's work, and their argument was that if Microsoft succeeded a precedent would be set in allowing larger companies to steal the core concepts of any software developer's work.

The court's judgment seemed to invalidate the copyrighting of the "look and feel" of an application. However, the judgement was based more on contract law rather than copyright law, as Microsoft and Apple had previously entered into a contract with respect to licensing of Apple's icons on Windows 1.0. Further, Apple had not acquired a software patent to protect the intellectual idea of the look and feel of its Macintosh operating system. Further, had Apple won then the development of X-Windows and other open-source GUIs would have been impeded.

7.3.2 The Browser Wars

The world wide web was invented by Tim Bernards Lee in the early 1990s. Microsoft was initially slow to respond to the rise of the internet. However, in the mid-1990s it expanded its product suite into the world wide web with Microsoft Network (MSN). This product was intended to compete directly against America On-Line (AOL). The company developed some key internet technologies such as ActiveX, VBScript and JScript. ActiveX is an application programming interface built on the Microsoft Component Object Model (COM), and VBScript and Jscript are script languages. The company also released a new version of Microsoft SQL Server that provided built-in support for internet applications. The company released a new version of Microsoft Office and Internet Explorer 4.0 (its internet browser) in 1997. This was the beginning of Microsoft dominance of the browser market. Netscape had dominated the market but as Internet Explorer 4.0 (and its successors) was provided as a standard part of the Windows operating system (and also on Apple computers) it led to the replacement of Netscape by Internet Explorer.

This led to a filing against Microsoft stating that Microsoft was engaged in anti-competitive practices by including the Internet Explore browser in the Windows operating system, and that Microsoft had violated an agreement signed in 1994. The court was asked to stop the anti-competitive practices of the company. The leaking of internal memos of the company on the internet caused a great deal of controversy in 1998. These documents went into detail of the threat that open source software posed to Microsoft and also mentioned possible legal action against Linux and other open source software.

Windows 2000 was released in early 2000 and included enhanced multimedia capabilities. The legal action taken by the US Department of Justice against Microsoft concluded in mid-2000, and the judgement called the company an "abusive monopoly". The judgment stated that the company should be split into two parts. However, this ruling was subsequently overturned on appeal. The company released Windows XP in 2001.

Microsoft launched its .NET initiative in 2002 and this is a new API for Windows programming. It includes a new programming language called C#. Microsoft also released a new version of the Visual Studio development product in 2002. Microsoft Vista is due to be released in 2007.

7.4 Motorola

Motorola[7] was originally founded as the Galvin Manufacturing Corporation in 1928. Paul Galvin and his brother Joseph purchased a battery eliminator business in Chicago. They incorporated Galvin Manufacturing Corporation later that year. The company initially had five employees and its first product was a battery eliminator. This was a device that allows battery-powered radios to run on standard household electric current. The company introduced one of the first commercially successful car radios in 1930. This was the Motorola model 5T71 radio and it sold for between $110 and $130. Paul Galvin created the brand name "Motorola" in 1930. The origin of the name is from "Motor" to highlight the company's new car radio, and the suffix "ola" was in common use for audio equipment at the time.

[7] This section is dedicated to the staff of the late Motorola plant in Blackrock, Cork, Ireland. Motorola set up a plant in Cork in the mid-1980s and at its peak it employed over 500 skilled software engineer and I was impressed by their professionalism and dedication to customer satisfaction. The plant developed the Operations and Maintenance Centre (OMC) which is a key part of the GSM system.

Motorola has come a long way since then and it is now a global leader in wireless, broadband and automotive communications technologies. It is internationally recognized for its innovation, excellence and its dedication to customer satisfaction. It has played a leading role in transforming the way in which people communicate, and the company has been a leader rather than a follower in technology. Its engineers have developed innovative products and services to connect people to each other.

Motorola's products have evolved over the years in response to changing customers' needs. Many of its products have been radio-related, starting with a battery eliminator for radios, to the first walkie-talkies, to cellular infrastructure equipment and mobile phones. The company was also strong in semiconductor technology,[8] including integrated circuits used in computers. This included the microprocessors used in the Apple Macintosh and Power Macintosh computers. The Power PC chip was developed in partnership with IBM. Motorola has a diverse line of communication products, including satellite systems and digital cable boxes.

The company is renowned for its dedication to quality and customer satisfaction. It defined and implemented the original Six Sigma™quality principles in 1983. Motorola was awarded the first Malcolm Baldridge National Quality Award granted by the US Department of Commerce in 1988 in recognition of its work on 6-sigma. The three commandments of six sigma philosophy are:

- The company needs to be focused on the Customer (and on Customer Satisfaction).
- Data must be gathered to provide visibility into performance with respect to the key goals and performance of the processes.
- Variation in processes needs to be eliminated (as variation leads to quality problems).

The use of Six Sigma by Jack Welsh within General Electric is well-known. Six-Sigma is a rigorous customer-focused, data-driven management approach to business improvement. Its objectives are to eliminate defects from every product and process, and to improve processes to do things better, faster and at a lower cost. It can be used to improve every activity and step of the business that is concerned with cost, timeliness and quality of results. It is designed to provide tangible business results directly related to the bottom line of the company. Motorola University provides training on six sigma.

7.4.1 Early Years

Motorola was founded as the Galvin Manufacturing Corporation in 1928 and its first product was the battery eliminator. It introduced one of the first commercially

---

[8] Motorola's semi-conductor product sector became a separate company (Freescale Semiconductor Inc.) in 2003, as Motorola decided to focus on its core activities following a major re-structuring of the company.

successful car radios in 1930. This was the 5T71 radio and it was installed in most automobiles. The company entered the field of mobile radio communications in 1936 with the development of the Motorola Police Cruiser mobile receiver. This product was preset to a single frequency and allowed vehicles to receive police broadcasts. Motorola's roots are in radio technology, and its core expertise in radio enabled it to become a leader in mobile phone communications in the mid-1980s.

The company entered the home radio business in 1937 and its products included phonographs (for playing recorded music) and home radios. It developed a lightweight two-way radio system in 1940 that was used for communication during the Second World War. It introduced its first commercial line of Motorola FM two-way radio systems and equipment in 1941.

It introduced its first television for the home entertainment business in 1947. This was the Golden View model VT71 and it was priced under $200. The Galvin Manufacturing Corporation officially became Motorola, Inc., in 1947 and it introduced its well known logo in 1955.

It established a research centre in Arizona to investigate the potential of new technologies such as the transistor. It would become one of the largest manufactures of semi-conductors in the world. However, its semi-conductor product sector became a separate company (Freescale Semiconductor Inc.) in 2003. Motorola's first mass-produced semiconductor was a transistor intended for car radios. The company introduced a pager in 1956 that allowed radio messages to be sent to a particular individual.

Motorola's radio equipment (including a radio transponder) was used by Neil Armstrong on the Apollo 11 lunar module for two-way voice communication on the moon. A Motorola FM radio receiver was used on NASA's lunar roving vehicle to provide a voice link between the Earth and the moon.

Motorola presented a prototype for the world's first portable telephone in 1973. The DynaTAC (Dynamic Adoptive Total Area Coverage) used a radio technology called cellular. The company would spend $100 million in the development of cellular technology, and commercial services of DynaTAC commenced in 1983. The first DynaTAC phone became available to consumers in 1984 and weighed almost 2 lbs.

Motorola made a strategic decision in 1974 to sell off its radio and television manufacturing division. The television manufacturing division produced the Quasar product line. Quasar was sold as a separate brand from Motorola, and all Motorola manufactured televisions were sold as "Quasar". The Quasar division was sold to Matsushita who were already well-known for the Panasonic television brand. The Matsushita acquisition of Motorola's Quasar division was the beginning of the end of the manufacturing of televisions by US companies.

Motorola televisions were transistorized coloured models that contained all of the serviceable parts in a drawer beneath the television. However, they had quality problems. The new Japanese management succeeded in producing televisions with significantly higher quality than Motorola and they had 5% of the defects of Motorola manufactured televisions. Motorola's executives later visited the Quasar plant near Chicago and were amazed at the quality and performance improvements.

The Japanese had employed the principles of total quality management based on Deming and Juran, and had focused on improvements to the process. This had led significant cost savings as less time was spent in reworking defective televisions.

The Japanese quality professionals had recognized that the cost of poor quality was considerable, and their strategy was to focus on the prevention of defects at their source. This led to a dramatic reduction in the number of defects, and a corresponding reduction in the costs of correcting the defects. The Motorola executives were amazed at the correlation between cost and quality, and this motivated the six-sigma quality improvement programme in Motorola.

There were allegations that the acquisition by Matsushita of Quasar was nothing more than a Japanese strategy to avoid paying tariffs on television sets imported into the United States. The "Quasar" brand was considered to be domestically made even though Quasar's engineering and manufacturing division was being scaled down. The Quasar televisions produced consisted of Japanese parts as the company moved away from engineering in the United States and focused on assembly and distribution.

7.4.2 Six-Sigma

The CEO of Motorola established the "Five Year Ten Fold Improvement Programme" as one of the top-10 goals of the company in 1981. This was a commitment by senior management to achieve significant quality and performance improvements to its products and services over the next 5 years.

The roots of six-sigma as a measurement standard goes back to the eighteenth century German mathematician Gauss. He introduced the concept of the normal distribution curve, and this curve is important in probability theory. It is also known as the Gaussian distribution or bell curve.

The curve has two parameters, and these are the mean (a measure of location or centre of the curve), and the standard deviation (a measure of variability from the mean). The mean is represented by the Greek letter μ (mu) and the standard deviation is represented by the Greek letter σ (sigma). The properties of the standard deviation is given in Table 7.1.

Walter Shewhart was one of the grandfathers of quality and he worked on quality improvements by reducing product variation in the 1920s. He demonstrated that

Table 7.1 Properties of sigma levels

σ-Level	Area of Curve within Sigma Level
1-sigma	68.27%
2-sigma	95.45%
3-sigma	99.73%
4-sigma	99.993%
5-sigma	99.99994%
6-sigma	99.9999998%

three sigma (or standard deviations) from the mean is the point where a process requires correction. The term "Six Sigma" was coined by a Motorola engineer called Bill Smith in the early 1980s, and it was used by Motorola both as a measure of process variability and as a methodology (Table 7.2) for performance and quality improvement. The application of the methodology by an organization leads to a change of culture in the organization.

Motorola engineers realized that the traditional quality levels of measuring defects in thousands of opportunities did not provide sufficient granularity for quality and performance improvements. The new Motorola standard allowed defects per million opportunities to be measured. Six-Sigma was a major success for Motorola, and the company made major savings following its introduction in the mid-1980s.

The objective of the 6σ programme was to improve the quality performance of all key operations in the business including manufacturing, service, marketing and support. The goal was to design and manufacture products that are 99.9997% perfect: i.e., 3.4 defects per million opportunities. The fundamental philosophy of the methodology is that every area of the organization can be improved. The steps involved in 6-sigma are summarized by the acronym DMAIC.[9] This stands for Define, Measure Analyse, Improve and Control. There is a step zero before you start and it is concerned with six-sigma leadership.

The methodology is based upon improving processes by understanding and controlling variation. This leads to more predictable processes with enhanced capability, and therefore more consistent results.

The participants on a six-sigma programme have an associated belt (e.g., green belt, black bet, etc.) to indicate their experience and expertise with the methodology. A Black Belt has received extensive training on six sigma and statistical techniques. Motorola was awarded the first Malcolm Baldrige National Quality Award in recognition of its efforts in 6σ. This Award was established by the US Congress to recognize the pursuit of quality in American business.

The use of six-sigma is not a silver bullet to success for any organization. Motorola paid the price for totally misjudging the transition from the analog cellular market to digital cellular.

Table 7.2 Six sigma methodology

Activity	Description
Define	Define the process.
Measure	Measure the current performance of the process.
Analyse	Analyse the process to identify waste.
Improve	Improve the process.
Control	Measure the improvements made to the process and repeat the cycle.

[9] DMAIC was influenced by Demings "Plan, Do, Act, Check" cycle.

7.4.3 Cellular Technologies

The invention of the telephone by Graham Bell in the late nineteenth century was a revolution in human communication, as it allowed people to communicate over distance. However, its key restriction was that the physical location of the person to be contacted was required before communication could take place: i.e., communication was between places rather than people.

Bell Laboratories introduced the idea of cellular communications in 1947 with the police car technology. Motorola was the first company to incorporate the technology into a portable device that was designed for use outside of an automobile. The inventor of the first modern portable handset was Martin Cooper of Motorola, and he made the first call on a portable cell phone in April 1973 to the head of research at Bell Labs.

It took a further 10 years for Motorola to bring the mobile phone to the market, and it introduced the first portable mobile phone in 1983. It was called DynaTAC, and it cost $3,500 and weighed one pound. Today, there are more mobile phone users than fixed line users, and mobile phones weigh as little as 3 ounces.

Bell Laboratories developed the Advance Mobile Phone Services (AMPS) standard for analog cellular phone service. It used the 800 MHz cellular band, and it was used in the United States and other countries. It had a frequency range between 800 and 900 MHz. These frequencies were allocated by the Federal Communications Commission (FCC). Each service provider could use half of the 824–849 MHz range for receiving signals from cellular phones and half the 869–894 MHz range for transmitting to cellular phones. The bands are divided into 30 kHz sub-bands called channels. The division of the spectrum into sub-band channels is achieved by using frequency division multiple access (FDMA).

The signals from a transmitter cover an area called a cell. As a user moves from one cell into a new cell a handover to the new cell takes place without any noticeable transition. The signals in the adjacent cell are sent and received on different channels than the previous cell's signals and so there is no interference. Analog is the original mobile phone system but has been replaced by more sophisticated systems in recent years. These include GSM, CDMA, GPRS and UMTS. The old analog phones were susceptible to noise and static. They were also subject to eavesdropping.

Bell Labs constructed and operated a prototype cellular system in Chicago in the late 1970s and performed public trials in Chicago in 1979. Motorola commenced a second US cellular radio-phone system test in the Washington/Baltimore area. The first commercial systems commenced operation in the United States in 1983.

Motorola dominated the analog mobile phone market. However, it was slow to adapt to the GSM standard and it paid the price in loss of market share to Nokia and other competitors. The company was very slow to see the potential of a mobile phone as a fashion device.[10] However, in recent years the company has made painful

---

[10] The attitude of Motorola at the time seemed to be similar to that of Henry Ford: i.e., they can have whatever colour they like as long as it is black.

adjustments to position itself appropriately for the future. The company decided to focus on mobile phone technology and has sold off non-core businesses including its automotive and semi-conductor businesses. It has also become much more customer focused and has launched a series of stylish phones.

7.4.4 Semiconductor Sector

Bell Labs invented the transistor in 1947 and its first commercial use was products for the hearing impaired. It invented an all-transistor computer in 1954. Motorola set up a research lab in 1952 to take advantage of the potential of semi-conductors, and by 1961 it was mass producing semi-conductors at a low cost. It introduced a transistorized walkie-talkie in 1962 as well as transistors for the Quasar televisions. It became the main supplier for the microprocessors used in Apple Macintosh and Power Macintosh personal computers. The power PC chip was developed in a partnership with IBM and Apple.

Motorola introduced the 8-bit 6,800 microprocessors in 1974 and this microprocessor was used in automotive, computing and video games. It contained over 4,000 transistors. It introduced a 16-bit microprocessor in 1979 and this was adopted by Apple for its Macintosh personal computers. It introduced the MC68020, the first true 32-bit microprocessor in 1984. This microprocessor contained 200,000 transistors on a 3/8 inch square chip.

Motorola went through a painful adjustment in the late 1990s and decided to focus on its core business of mobile communications. Its semiconductor business became a separate company called Freescale in 2004.

7.4.5 Motorola and Iridium

Iridium was launched in late 1998 to provide global satellite voice and data coverage to its customers. It provides complete coverage of the earth, and this includes the oceans, airways and polar regions. No other form of communication is available in remote areas, and so the concept of Iridium is very valuable.

Iridium was implemented by a constellation of 66 satellites. The original design of Iridium required 77 satellites, and the name "Iridium" was chosen as its atomic number in the periodic table is 77. However, the later design required 66 satellites, and so Dysprosium which has an atomic number 66 would now be a more appropriate name. The satellites are in low Earth orbit at a height of approximately 485 miles, and communication between satellites is via intersatellite links. The satellite contains seven Motorola Power PC 603E processors running at 200 MHz. These machines are used for satellite communication and control.

Iridium routes phone calls through space and there are four earth stations. As satellites leave the area of an Earth base station the routing tables change, and frames are forwarded to the next satellite just coming into view of the Earth base station.

The Iridium constellation is the largest commercial satellite constellation in the world, and is especially suited for industries such as maritime, aviation, government and the military. Motorola was the prime contractor for Iridium, and it played a key role in the design and development of the system. The satellites were produced at an incredibly low cost of $5 million each ($40 million each including launch costs).

The first Iridium call was made by Al Gore. However, Iridium as a company failed and it went into bankruptcy protection in late 1999 due to:

- insufficient demand for its services
- High cost of its service
- Cost of its mobile handsets
- Bulky mobile handsets
- Competition from the terrestrial mobile phone networks
- Management failures.

However, the Iridium satellites remained in orbit, and the service was re-established in 2001 by the newly founded Iridium Satellite LLC. It is used extensively by the US Department of Defence.

7.5 Apple Computers

Apple was founded by Steven Wozniak and Steven Jobs (Fig.7.10) in 1976. Jobs and Wozniak were two college dropouts, and they released the Apple I computer in 1977. It retailed for $666.66. They then proceeded to develop the Apple II computer. This machine included colour graphics and it came in its own plastic casing. It retailed for $1299 and it was one of the first computers to come pre-assembled. The Apple II was a commercial success and Apple became a public listed company in 1980.

The Apple Macintosh was released in 1984 and this machine was quite different from the IBM PC in that it included a graphical user interface that was friendly and intuitive. This made the Macintosh a much easier computer to use than the standard IBM PC, as the latter required users to be familiar with its DOS operating system and commands. The introduction of the Mac GUI was an important milestone in the computing field.

Jobs got the idea of the graphical user interface for the Macintosh from Xerox's PARC research centre in Palo Alto in California. Apple intended that the Macintosh would be an inexpensive and user friendly personal computer that would rival the IBM PC and its clones. However, it was more expensive than its rivals as it retailed for $2,495, and initially the Macintosh had limited applications available. The IBM PC had spreadsheets, word processors and databases applications available.

Apple went through financial difficulty in the mid-1980s as its products were more expensive than rival offerings. Jobs resigned from the company in 1985, and he founded a new company called Next Inc. Microsoft and Apple signed a contract in the mid-1980s that granted Microsoft permission to use some of the Macintosh GUIs. This contract would lead to Apple losing all of the lawsuits over copyright

Fig. 7.10 Steve Jobs
Photo courtesy of Wikipedia.

infringement against Microsoft that it took in later years. Apple Computers was sued by the Apple Corporation[11] in 1989 for violating the terms of their 1981 agreement that prohibited Apple Computers from building computers with the capability of producing synthesized music.

Apple released the Newton Message Personal Digital Assistant (PAD) in 1993. However, the Newton proved to be unsuccessful mainly due to reliability problems. IBM, Apple and Motorola entered an alliance in the early 1990s aimed at challenging the Windows and Intel architecture. The responsibilities of IBM and Motorola were to design and produce the new Power CPUs for the new desktop computers

---

[11] This was the Beatles recording company.

and servers. The responsibilities of Apple were to port its MacOS operating system to the new architecture.

The first Power PC processor was released by Motorola in 1993, and Apple released its power Macintosh desktop computer in 1994 based on the Motorola processors. Motorola developed processors with enhanced capabilities in the coming years, and this led to better and faster Power Macintosh computers. These include the Power Macintosh and PowerBook G3, G4 and G5. Apple took over Next Inc. in 1996, and this led to a return of Jobs to Apple. Jobs became the acting head of Apple in 1997 and he developed an alliance between Apple and Microsoft. The iMac was released in 1998 and this was a major commercial success. Apple released the iBook in 1999 and this was another major commercial success.

Apple released the iPod in 2001. This was a portable hard-disk MP3 player, and it had a capacity of 5 GB. The iPod could hold up to 1,000 MP3 songs. Apple released a software package (iTunes) that allowed MP3 files to be transferred from the Mac to the iPod.

Apple is a highly innovative company and has made major contributions to the history of computing.

7.6 Oracle

Oracle was founded in 1977 by Larry Ellison, Bob Miner and Ed Oates. Ellison came across a working prototype of a relational database, and he saw an opportunity to exploit and commercialise relational database technology. At that time there were no relational databases in the world, and the founding of Oracle changed business computing. The company was originally called Software Development Laboratories, and it changed its name to Oracle in 1983. The database product that Ellison, Milner and Oates created was called Oracle in memory of a CIA funded project code named Oracle that they had worked on in a previous company.

Today, Oracle is the main standard for database technology and applications, and Oracle databases are used by companies throughout the world. Oracle has grown over time to become the second largest software company[12] in the world. An Oracle database consists of a collection of data managed by an Oracle database management system. The release of Oracle V.2 in 1979 was a major milestone in the history of computing as it was the world's first relational database.

The concept of a relational database was described by Edgar Codd [Cod:70]. Codd was born in England and he worked for IBM in the United States. A relational database is a database that conforms to the relational model, and it may also be defined as a set of relations (or tables). A Relational Database Management System (RDBMS) is a system that manages data using the relational model, and examples include products such as Oracle and Microsoft SQL Server.

---

[12] Microsoft is the largest software company in the world.

A relation is defined as a set of tuples and is usually represented by a table, where a table is data organized in rows and columns. The data stored in a each column of the table is of the same data type. Constraints may be employed to provide restrictions on the kinds of data that may be stored in the relations. Constraints are Boolean expressions which indicate whether the constraint holds or not, and are a way of implementing business rules into the database.

Most relations have one or more key associated with them, and the key uniquely identifies the row of the table. An index is a way of providing quicker access to the data in a relational database. It allows the tuple in a relation to be looked up directly (using the index) rather than checking all of the tuples in the relation.

A stored procedure is executable code that is associated with the database. It is usually written in an imperative programming language, and it is used to perform common operations on the database.

Oracle is recognised as a world leader in relational database technology and its products play a key role in business computing.

7.7 Siemens

Siemens is a European technology giant and it employs approximately 475,000 people around the world. It is one of the world's leading electrical engineering and electronics companies, and its headquarters are in Munich and Berlin. Its annual sales are in excess of € 87 billion. The company operates in several segments including automation and control, power, transportation, medical, and information and communications. It was founded in 1847 by Werner von Siemens, and initially the focus of the company was in telecommunications with a product that improved upon the Wheatstone Telegraph. The company was involved in building telegraph networks in Prussia and Russia, as well as building an Indo-European[13] telegraph network in its early years. The company began laying the first transatlantic cable from Ireland to the United States in 1874. Siemens played a role in the introduction of facsimile telegraphy in the 1920s. This involved the scanning of photographs and it was employed on Siemens equipment from the late 1920s. It proved to be very popular with the press as it made it easier to transmit photographs. Siemens played a role in launching the world's first public telex network in work that it did with Deutsche Reichpost in 1933. This was achieved by technology to type and receive messages.

Siemens became the first company to succeed in manufacturing the ultra-pure silicon needed for semiconductor components. This was done in 1953 and was independent of the research being done in the United States. This technique was known as the floating-zone method, and the company was granted a patent for this

[13] The construction took 2 years and the route was over 11,000 km. It extended from London to Calcutta. It took 1 minute for a dispatch to reach Teheran, whereas it took 28 minutes to reach Calcutta.

invention in 1953. The company became involved in satellite communication for telecommunications from the mid-1960s.

Siemens has played an important role in the development of mobile phone technology. The company entered the mobile phone market in the early 1980s, and at that time there was a very expensive public mobile phone network[14] in place in Germany. Siemens supplied new mobile radio system equipment to Deutch Bundespost in the mid-1980s, and this technology allowed subscribers to be reached at their own numbers. It introduced its first mobile phone (C1) in the mid-1980s. It weighed 8.8 Kg or 19 pounds. Today, many mobile phones weigh less than 200 grams. Siemens made its first GSM call in the early 1990s.

Siemens has a number of joint ventures with companies such as Fujitsu, Nokia and Bosch. The joint venture with Fujitsu is now Europe's leading IT manufacturer. Fujitsu Siemens' portfolio ranges from high-performance servers, to PCs, notebooks, and so on. Nokia–Siemens Networks has world class research and development, and its mission is to advance the development of product platforms and services for the next-generation of fixed and mobile networks. Siemens is an innovative company and its products and services are familiar to consumers world-wide.

7.8 HP

Hewlett Packard was founded by Bill Hewlett and Dave Packard in 1939. They were both classmates at Stanford University, and graduated in engineering in 1934. Packard then took a position with General Electric, and Hewlett continued with graduate studies in Stanford/MIT. They built their first product in a Palo Alto garage, and this was an electronic test instrument used by sound engineers. Walt Disney Studios was an early HP customer.

The company began to grow during the early 1940s with orders from the US government. Hewlett and Packard created a management style for the company, and over time this became known as the HP way. The HP way was highly effective, and HP's corporate culture was later copied by several technology companies.

The HP way included a strong commitment by the company to its employees, and a strong belief in the basic goodness of people and in their desire to do a good job. It believed that if employees are given the proper tools to do their job that they would then do a good job. There was a firm conviction that each person had the right to be treated with respect and dignity.

The HP management technique was known as "management by walking around", and this is characterised by the personal involvement of management. This includes good listening skills by the manager, and the recognition that everyone in a company wants to do a good job. The HP way involves management by objectives: i.e., senior

[14] This network supported 11,000 users in Germany and was staffed by 600 staff who were responsible for switching the calls of the subscribers. The use of this network was the preserve of the super rich in Germany as the cost of phones and calls were prohibitive for the general public.

managers communicate the overall objectives clearly to their employees. Employees are then given the flexibility to work towards those goals in ways that are best for their own area of responsibility. The HP Way was refined further in the late 1950s, and the company objectives included seven areas. These are profit, customers, fields of interest, growth, people, management and citizenship.

HP also established an open door policy to create an atmosphere of trust and mutual understanding. The open door policy encouraged employees to discuss problems with a manager without fear of reprisals or adverse consequences. HP addressed employees by their first name and provided good benefits to its employees. This included free medical insurance and the provision of regular parties for employees. HP was the first company to introduce flexitime, and this was introduced in the late 1960s. The concept was based on trust and it allowed employees to arrive early or late for work as long as they worked a standard number of hours.

HP entered the microwave field during the Second World War and it later became a leader in the technology. It became a public quoted company in 1957, and set up two plants in Europe in the late-1950s.

It moved into the medical device sector in the 1960s and developed its first computer in the mid-1960s. HP's research laboratory was established in 1966, and it became a leading commercial research centre.

The company introduced the hand held scientific calculator in the early 1970s, and this invention made the slide rule obsolete. HP become a major player in the computer industry in the 1980s, and its product portfolio included desktop computers and mini-computers. It also introduced a set of inkjet and laser printers. It introduced a touchscreen personal computer in the early 1980s.

HP merged with Compaq in 2002, and the new company is known as HP. Today, HP has revenues of over $90 billion and employs over 150,000 people. The rise of HP and insight into its business practices, culture and management style that led to its success are described by David Packard in [Pac:96].

7.9 Miscellaneous

This section considers some miscellaneous technology companies including Philips, Amdahl and Sun Microsystems.

7.9.1 Amdahl

Amdahl was founded by Gene Amdahl in 1970. Amdahl was a former IBM employee and had worked on the System 360 family of mainframe computers. The company launched its first product in 1975, and this was the Amdahl 470. This product competed directly against the IBM System 370 family of mainframes. Amdahl became a major competitor to IBM in the high-end mainframe market, and Amdahl gained 24% market share.

Amdahl worked closely with Fujitsu to improve circuit design, and Fujitsu's influence on the company increased following Gene Amdahl's departure from the company in 1980. Amdahl moved into large system multi-processor design from the mid-1980s. However, by the late-1990s it was clear that Amdahl could no longer effectively compete against IBM's 64-bit zSeries as Amdahl had only 31-bit servers to sell. The company estimated that it would take $1 billion to create an IBM-compatible 64-bit system.

Amdah became a wholly owned subsidiary of Fujitsu in 1997, and its headquarters are in California.

7.9.2 Philips

Philips is a European technology giant and it was founded in Eindhoven, The Netherland in 1891. It was founded by Gerard Philips,[15] and initially the company made carbon-filament lamps. Today, it is a technology giant and it employs over 120,000 people, and has sales of approximately $27 billion. Its headquarters moved to Amsterdam in the late 1990s.

The company began manufacturing vacuum tubes and manufacturing radios in the 1920s. It introduced consumer products such as the electric razor in the 1930s, and introduced the compact cassette in the 1960s. By the mid-1960s it was producing integrated circuits, and its semiconductor sector played a key role in its business. Philips introduced the laser disc player in the late 1970s and it introduced the compact disk in 1982. It introduced the DVD in the late 1990s.

Philips sold off a majority stake of its semiconductor business to a consortium of private equity investors in 2005. The company has gone through a major process of change, and it plans to focus on health care, lifestyle and technology in the future.

7.9.3 Sun Microsystems

Sun Microsystems was founded by four Stanford University graduates in 1982, and the company is a vendor of computers and software. Its headquarters are in Santa Clara in California.

Sun's products include computer servers and workstations based on its SPARC processors and Solaris operating system. The SPARC workstation was first introduced in the late 1980s. It has developed innovative technologies such as the Java platform, and has contributed to the development of open systems and open source software. The Java technology plays a key role in the portability of software, and allows developers to write applications once to run on any computer.

---

[15] Gerard Philips was a cousin of Karl Marx.

7.10 Review Questions

1. Discuss the contribution of IBM to computing.
2. Discuss the contribution of Motorola to mobile phone technology.
3. Discuss the controversy between Microsoft and Apple and the controversy between Microsoft and Netscape.
4. Describe the 6σ methodology.
5. Describe the HP Way.

7.11 Summary

This chapter considered the history of some famous technology companies including IBM, Microsoft and Motorola.

The origin of IBM goes back to Hermann Hollerith's work on tabulating machines. This machine was designed to tabulate the results of the 1890 census in the United States.

Microsoft was founded in the mid-1970s and it has grown to become a major corporation. It has developed operating systems such as Microsoft Windows NT, Windows 2000 and Windows XP.

Motorola was founded as the Galvin Manufacturing Corporation in 1928. Its initial business was in the production of radios for cars and it became a world leader in radio and telecommunications. Motorola produces mobile phones and base stations for the mobile telecommunications field.

Apple was founded in the mid-1970s. It has developed innovative technology including the Apple Macintosh in the mid-1980s. This introduced a friendly graphical user interface (GUI) which made the machine easy to use.

Other companies discussed in this chapter include HP, Oracle and Siemens.

References

[AnL:95] The Heritage of Thales. W.S. Anglin and J. Lambek. Springer Verlag, New York, 1995.

[Bab:42] Sketch of the Analytic Engine. Invented by C. Babbage. L.F. Menabrea. Bibliothèque Universelle de Genève. Translated by Lada Ada Lovelace, 1842.

[Bar:69] G. Boole: A Miscellany. Edited by P.D. Barry. Cork University Press, 1969.

[Bec:00] Extreme Programming Explained. Kent Beck. Addison Wesley. 2000.

[Ber:99] Principles of Human Knowledge. G. Berkeley. Oxford University Press, 1999. (Originally published in 1710).

[BjJ:78] The Vienna Development Method: The Meta-Language. D. Bjørner and C. Jones. Lecture Notes in Computer Science, Vol. 61. Springer Verlag, 1978.

[BjJ:82] Formal Specification and Software Development. D. Bjørner and C. Jones. Prentice Hall International Series in Computer Science, 1982.

[BL:00] Weaving the Web. T. Berners-Lee. Collins Book, 2000.

[Boe:88] A Spiral Model for Software Development and Enhancement. B. Boehm. *Computer*, 21(5):61–72, May 1988.

[Boo:48] The Calculus of Logic. G. Boole. *Cambridge and Dublin Mathematical Journal*, 3:183–98, 1848.

[Boo:58] An Investigation into the Laws of Thought. G. Boole. Dover Publications, 1958. (First published in 1854).

[BoM:85] Program Verification. R.S. Boyer and J.S. Moore. Journal of Automated Reasoning. Vol 1(1). 1985.

[Bou:94] Formalisation of Properties for Feature Interaction Detection. Wiet Bouma et al. IS&N Conference. Springer Verlag. 1994.

[Brk:75] The Mythical Man Month. F. Brooks. Addison Wesley, 1975.

[Brk:86] No Silver Bullet: Essence and Accidents of Software Engineering. F.P. Brooks. Information Processing. Elsevier, Amsterdam, 1986.

[Bro:90] Rational for the development of the U.K. Defence Standards for Safety Critical Software. Compass Conference, 1990.

[Bus:45] As We May Think. V. Bush. *The Atlantic Monthly*, 176(1):101–108, July 1945.

[ChR:02] The Role of the Business Model in Capturing Value from Innovation: Evidence from Xerox Corporation's Technology Spin-off Companies. H. Chesbrough and R. Rosenbloom. *Industrial and Corporate Change*, 11(3):529–555, 2002.

[ClN:02] Software Product Lines. Practices and Patterns. SEI Series in Software Engineering. Paul Clements and Linda Northrop. Addison-Wesley. 2002.

[CKS:03] CMMI: Guidelines for Process Integration and Product Improvement. M.B. Chrissis, M. Conrad and S. Shrum. SEI Series in Software Engineering. Addison Wesley, 2003.

[Cod:70] A Relational Model of Data for Large Shared Data Banks. E.F. Codd. *Communications of the ACM*, 13(6):377–387, 1970.

[Crs:79] Quality is Free: The Art of Making Quality Certain. P. Crosby. McGraw Hill, 1979.

[Dem:86] Out of Crisis. W.E. Deming. M.I.T. Press, 1986.

[Des:99] Discourse on Method and Meditations on First Philosophy (4th edition). R. Descartes. Translated by D. Cress. Hackett Publishing Company, 1999.

[Dij:68] Go To Statement Considered Harmful. E.W. Dijkstra. *Communications of the ACM*, 11(3):147–148, March 1968

[Dij:72] Structured Programming. E.W. Dijkstra. Academic Press, 1972.

[Dij:76] A Disciple of Programming. E.W. Dijkstra. Prentice Hall, 1976.

[Fag:76] Design and Code Inspections to Reduce Errors in Software Development. M. Fagan. *IBM Systems Journal*, 15(3):182–211, 1976.

[Fen:95] Software Metrics: A Rigorous Approach. N. Fenton. Thompson Computer Press, 1995.

[Flo:63] Syntactic Analysis and Operator Precedence. R. Floyd. *Journal of the Association for Computing Machinery*, 10:316–333, 1963.

[Flo:64] The Syntax of Programming Languages: A Survey. R. Floyd. *IEEE Transactions on Electronic Computers*, EC-13(4):346–353, 1964.

[Flo:67] Assigning Meanings to Programs. R. Floyd. *Proceedings of Symposia in Applied Mathematics*, 19:19–32, 1967.

[Ger:02] Risk Based E-Business Testing. P. Gerrard and N. Thompson. Artech House Publishers, 2002.

[Ger:94] Experience with Formal Methods in Critical Systems. Susan Gerhart, Dan Creighton and Ted Ralston. IEEE Software, Jan '94.

[Geo:91] The RAISE Specification Language. A Tutorial. Chris George. Lecture Notes in Computer Science (552). Springer Verlag. 1991.

[Glb:94] Software Inspections. T. Gilb and D. Graham. Addison Wesley, 1994.

[Glb:76] Software Metrics. T. Gilb. Winthrop Publishers, Inc. Cambridge, 1976.

[Goe:31] Undecidable Propositions in Arithmetic. K. Goedel. *Über formal unentscheidbare Sätze der Principia Mathematica und verwandter Systeme, I. Monatshefte für Mathematik und Physik*, 38:173–198, 1931.

[Gri:81] The Science of Programming. D. Gries. Springer Verlag, Berlin, 1981.

[HB:95] Applications of Formal Methods. Edited by M. Hinchey and J. Bowen. Prentice Hall International Series in Computer Science, 1995.

[Hea:56] Euclid: The Thirteen Books of the Elements (Vol. 1). Translated by Sir T. Heath. Dover Publications, New York, 1956. (First published in 1925).

[Hor:69] An Axiomatic Basis for Computer Programming. C.A.R. Hoare. *Communications of the ACM*, 12(10):576–585, 1969.

[Hor:85] Communicating Sequential Processes. C.A.R. Hoare. Prentice Hall International Series in Computer Science, 1985.

[HoU:79] Introduction to Automata Theory, Languages and Computation. John E. Hopcroft and Jeffrey D. Ullman. Addison-Wesley. 1979.

[Hum:06] An Enquiry Concerning Human Understanding (Paperback). D. Hume. Digireads.com, 2006. (Originally published in 1748).

[Hum:89] Managing the Software Process. W. Humphry. Addison Wesley, 1989.

[Jac:04] The Unified Modeling Language Reference Manual (2nd Edition). Grady Booch, James Rumbaugh and Ivar Jacobson, Addison Wesley. 2004.

[Jur:00] Juran's Quality Handbook (5th edition). J. Juran. McGraw Hill, 2000.

[KaC:74] Protocol for Packet Network Interconnections. B. Kahn and V. Cerf. *IEEE Transactions on Communications Technology*, May 1974.

[Kan:03] Critique of Pure Reason. Immanuel Kant. Dover Publications. 2003.

[KeR:78] The C Programming Language (1st edition). B. Kernighan and D. Ritchie. Prentice Hall Software Series, 1978.

[Ker:81] Why Pascal is Not My Favourite Language. B. Kernighan. AT&T Bell Laboratories, 1981.

[KeR:88] The C Programming Language (2nd edition). B. Kernighan and D. Ritchie. Prentice Hall Software Series, 1988.

[Knu:97] The Art of Computer Programming (Vol. 1, 3rd edition). D. Knuth. Addison Wesley, 1997.

[Kuh:70] The Structure of Scientific Revolutions. T. Kuhn. University of Chicago Press, 1970.

[Lak:76] Proof and Refutations: The Logic of Mathematical Discovery. I. Lakatos. Cambridge University Press, 1976.

[Lon:58] Insight: A Study of Human Understanding. B. Lonergan. Longmans, Green and Co, 1958.

[MaP:02] Boo Hoo: $135 Million, 18 Months ... A Dot.Com Story from Concept to Catastrophe. E. Malmsten and E. Portanger. Arrow, 2002.

[Mac:90] Conceptual Models and Computing. PhD Thesis. Dept. of Computer Science. Trinity College Dublin. 1990.

[Mac:93] Formal Methods and Testing. Mchel Mac An Airchinnigh. Tutorial of 6th International Software Quality Week. Software Research Institute. San Francisco. 1993.

[Mar:82] Applications Development Without Programmers. J. Martin. Prentice Hall, 1982.

[Mc:59] Programs with Common Sense. J. McCarthy. *Proceedings of the Teddington Conference on the Mechanization of Thought Processes*, Her Majesty's Stationary Office, 1959.

[McD:94] MSc Thesis. Dept. of Computer Science. Trinity College Dublin. 1994.

[McH:85] Boole. D. McHale. Cork University Press, 1985.

[McP:43] A Logical Calculus of the Ideas Immanent in Nervous Activity. W. McCullock and W. Pitts. *Bulletin of Mathematics Biophysics*, 5:115–133, 1943.

[Men:87] Introduction to Mathematical Logic. E. Mendelson. Wadsworth and Cole/Brook Advanced Books & Software, 1987.

[Mil:89] Communication and Concurrency. Robin Milner. International Series in Computer Science. Prentice Hall. 1989.

[MOD:91a] Def. Stan 00-55. Requirements for Safety Critical Software in Defence Equipment. Interim Defence Standards U.K., 1991.

[MOD:91b] Def. Stan 00-56. Guidance for Safety Critical Software in Defence Equipment. Interim Defence Standards U.K., 1991.

[Nau:60] Report on the Algorithmic Language: ALGOL 60. Edited by P. Naur. *Communication of the ACM*, 3(5):299–314, 1960.

[NeS:56] The Logic Theory Machine. A. Newell and H. Simon. *IRE Transactions on Information Theory*, 2:61–79, 1956.

[Nie:01] Web Design in a Nutshell: A Desktop Quick Reference (2nd edition). J. Niederst. O'Reilly Publications, 2001.

[ONu:95] The Search for Mind. S. O'Nuallain. Ablex Publishing, 1995.

[ORg:97] Modelling Organisations and Structures in the Real World. G. O'Regan. Ph.D. Thesis. Trinity College Dublin, 1997.

[ORg:02] A Practical Approach to Software Quality. G. O'Regan. Springer Verlag, New York, 2002.

[ORg:06] Mathematical Approaches to Software Quality. G. O'Regan. Springer Verlag, London, 2006.

[Pac:96] The HP Way: How Bill Hewlett and I Built Our Company. D. Packard. Collins Press, 1996.

[Pau:93] Key Practices of the Capability Maturity Model. Mark Paulk et al. Software Engineering Institute. 1993.

[Plo:81] A Structural Approach to Operational Semantics. G. Plotkin. Technical Report DAIM FN-19. Computer Science Department, Aarhus University, Denmark, 1981.

[Pol:57] How to Solve It: A New Aspect of Mathematical Method. G. Polya. Princeton University Press, 1957.

[Pop:97] The Single Transferable Voting System. Michael Poppleton. IWFM97. Editors Gerard O'Regan and Sharon Flynn. Springer Verlag. 1997.

[Por:98] Competitive Advantage: Creating and Sustaining Superior Performance. M.E. Porter. Free Press, 1998.

[Pri:59] An Ancient Greek Computer. D.J. de Solla Price. Implicit in Scientific American, pp. 60–67, June 1959.

[Res:84] Mathematics in Civilization. H.L. Resnikoff and R.O. Wells. Dover Publications, New York, 1984.

[Roy:70] Managing the Development of Large Software Systems. Winston Royce. Proceedings of IEEE WESTCON (26). 1970.

[RuW:10] Principia Mathematica. B. Russell and A. Whitehead. Cambridge University Press, 1910.

[Sea:80] Minds, Brains, and Programs. J. Searle. *The Behavioral and Brain Sciences*, 3:417–457, 1980.

[Sha:37] A Symbolic Analysis of Relay and Switching Circuits. C. Shannon. Masters Thesis. Massachusetts Institute of Technology, 1937.

[Sha:48] A Mathematical Theory of Communication. C. Shannon. *Bell System Technical Journal*, 27:379–423, 1948.

[Sha:49] Communication Theory of Secrecy Systems. *Bell System Technical Journal*, 28(4):656–715, 1949.

[Smi:23] History of Mathematics (Vol. 1). D.E. Smith. Dover Publications, New York, 1923.

[Spi:92] The Z Notation: A Reference Manual. J.M. Spivey. Prentice Hall International Series in Computer Science, 1992.

[Std:99] Estimating: Art or Science. Featuring Morotz Cost Expert. Standish Group Research Note, 1999.

[Tie:91] The Evolution of Def. Standard 00-55 and 00-56. An Intensification of the Formal Methods Debate in the U.K. Margaret Tierney. Research Centre for Social Sciences. University of Edinburgh, 1991.

[Tur:50] Computing, Machinery and Intelligence. A. Turing. *Mind*, 49:433–460, 1950.

[Turn:85] Miranda. D. Turner. Proceedings IFIP Conference, Nancy, France, LNCS (201), Springer, 1985.

[VN:32] *Mathematische Grundlagen der Quantenmechan* (The Mathematical Foundations of Quantum Mechanics). J. von Neumann. Springer, Berlin, 1932.

[VN:45] First Draft of a Report on the EDVAC. J. von Neumann. University of Pennsylvania, 1945.

[Wit:22] Tractatus Logico-Philosophicus. L. Wittgenstein. Kegan Paul, London, 1922.

[Wrd:92] Formal Methods with Z: A Practical Approach to Formal Methods in Engineering. J.B. Wordsworth. Addison Wesley, 1992.

Glossary

AI	Artificial Intelligence
AIX	Advanced IBM UNIX
AOL	America On-Line
AMN	Abstract Machine Notation
AMPS	Advanced Mobile Phone Services
ANS	Advanced Network Service
ANSII	American Standard for Information Interchange
APPN	Advanced Peer-To-Peer Networking
ARPA	Advanced Research Project Agency
AT&T	American Telephone and Telegraph Company
ATM	Automated Teller Machine
B2B	Business to Business
B2C	Business to Consumer
BBN	Bolt Beranek and Newman
BNF	Backus Naur Form
CCS	Calculus Communicating Systems
CDMA	Code Division Multiple Access
CERN	Conseil European Recherche Nucleaire
CERT	Computer Emergency Response Team
CGI	Common Gateway Interface
CICS	Customer Information Control System
CMM®	Capability Maturity Model
CMMI®	Capability Maturity Model Integration
COBOL	Common Business Oriented Language
CODASYL	Conference on Data Systems Languages
COM	Component Object Model
CP/M	Control Program for Microcomputers
CPU	Central Processing Unit
CSP	Communication Sequential Processes
CTR	Computing Tabulating Recording Company
DARPA	Defence Advanced Research Project Agency
DES	Data Encryption Standard
DNS	Domain Naming System

DOS	Disk Operating System
DRAM	Dynamic Random Access Memory
DSDM	Dynamic Systems Development Method
DMAIC	Define, Measure, Analyse, Improve, Control
EJB	Enterprise Java Beans
EBCDIC	Extended Binary Coded Decimal Interchange Code
ENIAC	Electronic Numerical Integrator and Computer
ESI	European Software Institute
FCC	Federal Communications Commission
FDMA	Frequency Division Multiple Access
FTP	File Transfer Protocol
GPRS	General Packet Radio Service
GSM	Global System for Mobile Communication
GUI	Graphical User Interface
HICL	High Integrity Computing Laboratory
HTML	Hypertext Markup Language
HTTP	Hyper Text Transport Protocol
IBM	International Business Machines
IEEE	Institute of Electrical and Electronic Engineers
IMP	Interface Message Processors
IP	Internet Protocol
IPO	Initial Public Offering
IPTO	Information Processing Technology Office
ISO	International Standards Organization
JAD	Joint Application Development
JVM	Java Virtual Machine
LAN	Local Area Network
LT	Logic Theorist
MIT	Massachusetts Institute of Technology
MOD	Ministry of Defence (U.K.)
MSN	Microsoft Network
NAP	Network Access Point
NCP	Network Control Protocol
NPL	National Physical Laboratory
NSF	National Science Foundation
NWG	Network Working Group
OSI	Open Systems Interconnection
OS/2	Operating System 2
PC	Personal Computer
PDA	Personal Digital Assistant
PIN	Personal Identification Number
PS/2	Personal System 2
RACE	Research Advanced Communications Europe
RAD	Rapid Application Development
RAISE	Rigorous Approach to Industrial Software Engineering

RDBMS	Relational Database Management System
RISC	Reduced Instruction Set Computer
RSL	RAISE Specification Language
RSRE	Royal Signals and Radar Establishment
SAA	Systems Application Architecture
SAGE	Semi-Automatic Ground Environment
SCORE	Service Creation in an Object Reuse Environment
SDL	Specification and Description Language
SECD	Stack, Environment, Core, Dump
SEI	Software Engineering Institute
SLA	Service Level Agreement
SMTP	Simple Mail Transfer Protocol
SNA	Systems Network Architecture
SPI	Software Process Improvement
SPICE	Software Process Improvement and Capability determination
SQA	Software Quality Assurance
SQRL	Software Quality Research Laboratory
SRI	Stanford Research Institute
SSL	Secure Socket Layer
TCP	Transport Control Protocol
UCLA	University of California (Los Angeles)
UDP	User Datagram Protocol
UML	Unified Modelling Language
UMTS	Universal Mobile Telecommunications System
URL	Universal Resource Locator
VDM	Vienna Development Method
VDM♣	Irish School of VDM
VM	Virtual Memory
W3C	World Wide Web Consortium
XP	eXtreme Programming

Index

Printed in the United States
127181LV00001B/255/P

9 781848 000834